SpringerBriefs in Applied Sciences and Technology

SpringerBriefs present concise summaries of cutting-edge research and practical applications across a wide spectrum of fields. Featuring compact volumes of 50 to 125 pages, the series covers a range of content from professional to academic.

Typical publications can be:

- A timely report of state-of-the art methods
- An introduction to or a manual for the application of mathematical or computer techniques
- A bridge between new research results, as published in journal articles
- A snapshot of a hot or emerging topic
- An in-depth case study
- A presentation of core concepts that students must understand in order to make independent contributions

SpringerBriefs are characterized by fast, global electronic dissemination, standard publishing contracts, standardized manuscript preparation and formatting guidelines, and expedited production schedules.

On the one hand, **SpringerBriefs in Applied Sciences and Technology** are devoted to the publication of fundamentals and applications within the different classical engineering disciplines as well as in interdisciplinary fields that recently emerged between these areas. On the other hand, as the boundary separating fundamental research and applied technology is more and more dissolving, this series is particularly open to trans-disciplinary topics between fundamental science and engineering.

Indexed by EI-Compendex, SCOPUS and Springerlink.

More information about this series at http://www.springer.com/series/8884

Saiful Bahri Mohamed · Radzuwan Ab Rashid
Martini Muhamad · Jailani Ismail

Down Milling Trimming Process Optimization for Carbon Fibre-Reinforced Plastic

Saiful Bahri Mohamed
Universiti Sultan Zainal Abidin
Kuala Terengganu, Malaysia

Radzuwan Ab Rashid
Universiti Sultan Zainal Abidin
Kuala Terengganu, Malaysia

Martini Muhamad
Universiti Sultan Zainal Abidin
Kuala Terengganu, Malaysia

Jailani Ismail
Universiti Sultan Zainal Abidin
Kuala Terengganu, Malaysia

ISSN 2191-530X   ISSN 2191-5318   (electronic)
SpringerBriefs in Applied Sciences and Technology
ISBN 978-981-13-1803-0   ISBN 978-981-13-1804-7   (eBook)
https://doi.org/10.1007/978-981-13-1804-7

Library of Congress Control Number: 2018949351

This Springer imprint is published by the registered company Springer Nature Singapore Pte Ltd.
The registered company address is: 152 Beach Road, #21-01/04 Gateway East, Singapore 189721, Singapore

Preface

This book discusses the down milling trimming process optimization for carbon fiber-reinforced plastic (CFRP). The use of advanced materials which are composed of carbon fiber, polymers, and metal is increasing due to their special mechanical and physical properties. These materials are used in the aeronautical, aerospace, automotive, biomechanical, mechanical, and other industries. As a result of these properties and potential applications, there exists an urgent need to understand questions associated with the machinability of these materials. It is common for two different materials to produce different outcomes even though they are cut with the same tool at the same cutting speed and feed rates by the same machine which works under similar conditions. Some materials may produce long curly chips (like mild steel); some may produce short chips (like cast iron); some may get a smooth finish; some may end up with a rough surface; some may produce chatter; and some may produce lots of heat and quickly blunt the tool. Carbon fiber-reinforced plastic with aluminum grade 2024 (CFRP/Al2024) composite materials integrated into one single machining operation has proven to be more challenging due to the aniso-tropic and non-homogeneous structure of CFRP and ductile nature of aluminum. It causes several types of damages, such as matrix cracking and thermal alterations, fiber pullout and fuzzing during drilling and trimming which affect the quality of the machined surface. These problems mainly occur due to inappropriate use of various cutting tool design materials and cutting parameters. The research project discussed in this book aims to study and model machined surface quality of CFRP/Al2024 using two-level full factorial design experiment. This research pro-ject has three objectives: first, to perform the trimming process using down milling; second, to statistically and graphically analyze the influence and interaction of cutting parameters; and third, to optimize cutting parameters in order to get the surface texture quality of CFRP/Al2024 to less than 1 μm. The trimming process was carried out via down milling on a stack of multidirectional CFRP/Al2024. Three cutting parameters were considered, namely spindle speed (N), feed rate (f_r), and depth of cut (d_c). Two-level full factorial design was utilized to plan systematic experimental methodology. The analysis of variance (ANOVA) was used to analyze the influence and the interaction factors associated with surface quality. The results showed that the depth of cut is the most significant factor for Al2024, and for CFRP, the spindle speed and feed rate are significant. The validation test showed average deviation of predicted to actual value surface roughness is 3.11% for CFRP and 3.43% for Al2024. Optimization of surface roughness for CFRP/Al2024 of below that 1 μm can be obtained at the setting of $N = 11750$ rpm, $fr = 750$ mm/min and $C = 0.255$ mm, respectively.

Kuala Terengganu, Malaysia

Saiful Bahri Mohamed
Radzuwan Ab Rashid
Martini Muhamad
Jailani Ismail

Contents

Chapter 1
Composite Materials and Types of Machining

1.1 Defining Composite Material

A composite material is a composition of two or more constituent materials to form an overall structure which significantly better in terms of mechanical or chemical properties than the sum of individual material. Savage defines composite materials as a material in which two or more constituents have been brought together to produce a new material consisting of at least two chemically distinct components, with resultant properties significantly different to those of the individual constituents. These so-called engineered materials are combined in such a way to produce a new material which may be preferred for many reasons. However, the primary reason of choosing composite materials for components is because of weight saving for its relative stiffness and strength when compared to traditional materials. For example, carbon fiber-reinforced composite can be five times stronger than 1020 grade steel while having one fifth of the weight.

The advanced composite materials use glass, carbon, and aramid (Kevlar) popular for applications which demand high strength and stiffness, or low thermal conductivity. Most of advanced composite substitutes several aerospace metal parts [1, 2, 3]. Carbon fiber-reinforced plastics (CFRPs) are used in many applications, which required thermal stability, high-temperature strength, good ablation characteristics, and insulating capability. Glass fiber-reinforced plastic (GFRP) fibers are used in places which required greater strength and higher thermal conductivity and have six times the tensile strength of carbon fibers.

Pop et al. [4] classified three types of composite material which are based on the nature of its matrix, namely composite materials with the organic matrix, metallic matrix, and ceramics matrix. The composite materials that are commonly used in the aerospace industry are carbon fiber-reinforced polymer (CFRP) and fiber metal laminates (FMLs). This is due to their strength-to-weight ratio and good mechanical properties [5, 6].

S. B. Mohamed et al., *Down Milling Trimming Process Optimization for Carbon Fiber-Reinforced Plastic*, SpringerBriefs in Applied Sciences and Technology, https://doi.org/10.1007/978-981-13-1804-7_1

1.2   CFRP Composite Laminates

Aircraft manufacturing and aerospace industries utilize CFRP composite laminates due to their attractive properties such as weight-to-strength ratio, durability, and extreme corrosion resistance [7, 8]. Fiber-reinforced composite laminates commonly used in industries are carbon fiber-reinforced polymer (CFRP) composite laminates, glass fiber-reinforced polymer (GFRP) composite laminates, and fiber metal composite laminates (FMLs). They are being used to replace conventional metallic materials in several industries including aerospace, aircraft, and defense, which require structural materials with superior properties such as high strength to weight and stiffness to weight [9]. CFRP composite laminates offer some advantages including weight reduction, high strength, and stiffness (easily formable material) [10, 11]. These materials can also increase payload and fuel efficiency.

Fibers are well known due to their lightweight, strong, and stiff where they are the major contributor to the stiffness and strength of composites' laminates. The polymer matrix such as thermoset and thermoplastic binds the fibers together then transferring the load to the reinforced fibers and protect the fibers from the environmental attack [11].

1.3   Fiber Metal Laminate (FML)

High-performance and lightweight structures in aircraft industry stimulate research and development in introducing refined model for fiber metal laminates (FMLs). FML recommends major improvements to the present existing materials for aircraft structure [12]. FMLs are multicomponent materials utilizing metals, fibers, and matrix resins. Typical FMLs are prepared by stacking alternating layers of metal foils and fiber/matrix resin prepreg followed by consolidation in a press or autoclave. FMLs combine the best properties of the metal and the composite and this makes FML suitable for aerospace applications.

FMLs are considered as hybrid composite materials built up from interlacing layers of thin metals, such as aluminum alloys and fiber-reinforced adhesives. These two combinations of metal and polymer composite can produce a synergistic effect on some properties such as physical and mechanical properties. FML offers excellent damage tolerance, fatigue, and impact properties with a relative low density [10]. In addition, it offers great mechanical properties, like high fatigue resistance, high strength, high fracture toughness, high impact resistance, and high-energy absorbing capacity [10].

1.4   Classification of Composite Machining

Machining process of composite material has several objectives. The first objective is to create holes, slots, pockets, and other complex features which are not acquired during the manufacturing of the parts or components [11]. The components or parts made of composite materials are usually manufactured by net shape and need the removal of excess material in order to create and control the desired tolerances of the component [13]. Machining operation especially in milling operation is utilized as a corrective operation to produce smoothness and better surface finish. The most important purpose of machining process of composite materials is to produce a quality finish product and reduce manufacturing cost. The machining process is performed to make prototype part from the big blank or sheet of materials. This process is very reasonable instead of making an expensive mold where the molds are not flexible and restricted to mold size only [11].

Machining of the composites differs significantly in various aspects from machining of conventional metals and their alloys. The composites' behavior is not only being non-homogenous and anisotropic but also conveys diverse reinforcement and matrix properties and the volume fraction of matric and reinforcement. The tool encounters alternatively matrix and reinforcement materials, whose response to machining can be utterly different [14]. Secondary processing of machining such as trimming to final shape and drilling is typically essential to facilitate component assembly as it exhibits surface roughness of the workpiece. Surface roughness and tolerance are closely related and it is generally necessary to specify a smooth finish to maintain a fine tolerance in the finishing process. For many practical design applications, tolerance and strength requirements impose a limit on the maximum allowable roughness [15]. Thus, machining of composite materials enforces special demands on the optimization of influence and interaction of machining parameters in order to minimize the defects of the surface produced by machining, which can drastically affect the strength and chemical resistance of the material.

Traditional methods of machining often induce critical flaws in the component parts during the net trimming and various degrees of delamination, splintering, fiber pullout, and cracking have been reported [15–17]. The abrasive water jet (AWJ) is one such method which appears to be highly suited for production trimming of RFP materials [18]. Owing to its ability to achieve a quality surface at rapid production rates, AWJ is currently being sought for production applications [19]. However, very little experimental results are known regarding the effects of AWJ machining on the surface integrity of FRPs, besides there is exposure of moisture to the fiber which is significant will be affecting the composites physically.

The temperature during the cutting process must be considered where it should not exceed the cure temperature or close to the melting temperature of the resin to avoid material disintegration [20]. During the machining process, the smaller values of feed rate are much better as far as the tool life is concerned. However, this causes the process to be more time consuming and expensive [21]. Machining also exposes the fiber to chemical and moisture and the effect of coolant materials on composites [22].

The machining of composite materials requires the dry machining process which does not use any coolant or cutting fluid. This is because composite materials will expand if the coolant or any cutting fluid is used during machining. The mechanical damage will occur faster when the workpiece that is manufactured from composites are exposed to moisture and elevated temperature. The presence of water modifies the properties of the matrix and affected the mechanical properties such as strength, stiffness, and creep. Furthermore, there are several advantages of using dry cutting, such as no air and water pollutions, no residue on the swarf which will reflect on reduced disposal and cleaning cost, no harms to health and allergy free, and more importantly, it will reduce the cost of manufacturing [23].

1.4.1 Drilling

The most common traditional method in machining composites is drilling. This process is very important to make holes for assembling the subcomponent during the processing stage. Delamination is the main failure in drilling of composite materials. Several studies have been carried out to investigate the damages during the drilling operation of CFRP. For instance, Krishnaraj et al. [24] analyzed the effect of drill parameters namely drill diameter, spindle speed, and feed rate in drilling of multi-material (CFRP/Al) using Taguchi method. They found that the feed rate and drill diameter are the most significant parameters to the overall performance during the drilling of CFRP/Al stack.

Zitoune et al. [25] carried out a study on drilling of composite material and aluminum stack (CFRP/Al2024). They studied on the influence of diameter, spindle speed, and feed rate on thrust force, torque, surface finish, hole diameter, and also the circularity. The experimental results showed that the quality of holes could be improved by proper selection of cutting parameters. Krishnaraj et al. [24] used Taguchi L_{27} orthogonal array to perform drilling of CFRP composite plates. Gray fuzzy optimization of drilling parameters was conducted based on five different output performance characteristics namely thrust force, torque, entry delamination, exit delamination, and eccentricity of the holes. They concluded that feed rate is the most significant factor in drilling of CFRP composites.

Shyha et al. [26] drilled 30-mm-thick titanium/CFRP/aluminum workpiece stack. A modified fractional factorial designed based on a L_{18} Taguchi orthogonal array was employed. They found that damage occurred when drilling through the titanium (Ti-6Al $= 4$ V) rather than the aluminum (Al7050) or CFRP (unidirectional "UD" laminates). Feed rate and environment condition were significant in relation to thrust force for both Ti and Al.

1.4.2 Turning

Turning has become an important process for finish machining of high accuracy parts and highly precise joint areas in composite machining. Therefore, deep understanding of the behavior of composite components in turning is required for its successful implementation. The machinability of composites in turning has been studied in terms of tool wear, cutting forces, cutting temperatures, and surface quality. Fiber type, orientation, and volume fraction are the most significant material properties that influence machinability [27].

Some research work has been carried out in applying turning processes of composite materials with different cutting tools. Turning differs from milling and sawing mainly because of the constant engagement of the tool exists. The machinability of FRP is primarily determined by the physical properties of the fibers and the matrix as well as by the fiber orientation and volume fraction [14].

Palanikumar et al. [28] conducted investigation on the optimization of machining parameters for surface roughness of glass fiber-reinforced plastics (GFRP) using the design of experiments (DoE). The machining parameters considered were speed, feed, depth of cut, and workpiece (fiber orientation) in the turning operation. The statistical technique, ANOVA, is employed to analyze the experimental results. They concluded that feed rate is the most significant cutting parameters for surface roughness followed by cutting speed while depth of cut contributes the least significant parameter.

Optimization of surface roughness in turning unidirectional glass fiber-reinforced plastics (UD-GFRP) composites using polycrystalline diamond (PCD) cutting tool was conducted by Kumar et al. [29]. The study investigated the effect of tool nose radius, tool rake angle, feed rate, cutting speed, depth of cut and along with cutting environment (dry, wet, and cooled (5–7 °C) temperature) on the surface roughness produced. The experimental results revealed that the most significant machining parameters for surface roughness are feed rate followed by cutting speed. Cutting environment does not influence the surface roughness significantly.

1.4.3 Milling

There are two distinct ways to cut materials when milling, conventional (up) milling and climb (down) milling. The difference between these two techniques is the relationship of the rotation of the cutter to the direction of feed [22].

In conventional milling, the cutter rotates against the direction of the feed while during the climb milling, and the cutter rotates with the feed as illustrated in Figs. 1.1 and 1.2.

Conventional milling is the traditional approach when cutting because the backlash , the play between the lead screw and the nut in the machine table, is eliminated.

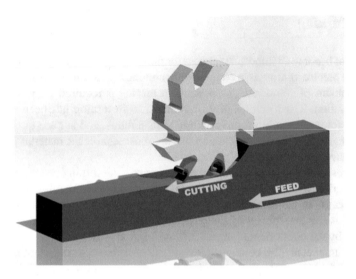

Fig. 1.1 Climb or down milling

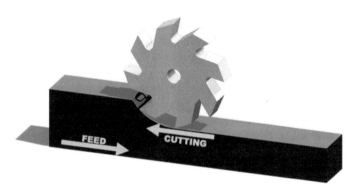

Fig. 1.2 Up milling

Recently, climb milling has been recognized as the preferred way to approach a workpiece due to the fact that more and more machines compensate for backlash or have a backlash eliminator.

Climb or down milling is generally the best way to machine parts today since it reduces the load from the cutting edge, leaves a better surface finish, and improves tool life. During the conventional milling, the cutter tends to dig into the workpiece and may cause the part to be cut out of tolerance. Even though the climb milling is the preferred way to machine parts, there are times when conventional milling is the recommended choice. Backlash, which is typically found in older and manual machines, is a huge concern with climb milling. If the machine does not counteract backlash, conventional milling should be implemented. Conventional milling is also suggested for use on casting or forgings or when the part is case hardened since the cut begins under the surface of the material.

Rahim et al. [5] carried out a project to improve the hole quality of CFRP/AL2024 in terms of surface roughness, cutting force, and temperature using helical milling technique. Helical milling was performed on a CFRP/Al2024 with a thickness of 7.6 and 3 mm, respectively. Both plates have the same dimension of 300 and 100 mm for length and width, respectively. The CFRP composite was made using an unidirectional based on aerospace requirement. The lay-up sequence is starting $0°$ and then oriented to $90°$, followed by $45°$ and $-45°$ angle. Three cutting parameters were examined, i.e., cutting speed, feed rate, and depth per helical path. They suggested that helical milling process, with combination of suitable machining parameters and cutting tool design can produce high quality of hole. They found that high feed rate and depth per helical cut produced high value of thrust force.

Karpat et al. [6] investigated cutting force model during the machining of CFRP laminates using slot milling and PCD cutters. The mechanistic force model and experimental results agreed with each other. They concluded that the model is capable to predict cutting forces.

Davim and Reis [13] performed evaluation on the cutting parameters (cutting velocity and feed rate) under the surface roughness and damage in milling laminate plates of CFRP. An analysis of variance (ANOVA) was performed to investigate the cutting characteristics of CFRP composite material using cemented carbide (K10) end mills. They concluded that the surface roughness (Ra) and international dimensional precision (IT) increased with feed rate and decreased with cutting velocity. Feed rate was the cutting parameter that presented the highest statistical and physical influence on surface roughness.

1.4.4 Trimming

Trimming is a process of correcting the boundary corner of materials, and it is a compulsory process once the panel of work material completed. Trimming is a part of milling process because the tool and machine used are still the same. However in trimming process, machinist also uses burr tool or router to cut the rough edges of the workpiece using a milling machine. The position of cutting tool perpendicularly relative to the workpiece will distinguish between milling and trimming processes. Figure 1.3 presents up trimming process, and Fig. 1.4 presents down trimming process.

Haddad et al. [30, 31] studied the trimming damages on CFRP structures and the impact of mechanical behavior. The composite specimens were prepared using three cutting processes: (i) an abrasive water jet (AWJ), (ii) a diamond cutter (ADS), and (iii) a standard cutting tool. The standard cutting tool machining experiments were performed using a DUBUS three-axis milling machine. These experiments were conducted using a full experimental design with three cutting speeds (350, 700, and 1400 m/min) and three feed speeds (125, 250, and 500 mm/min). However, this

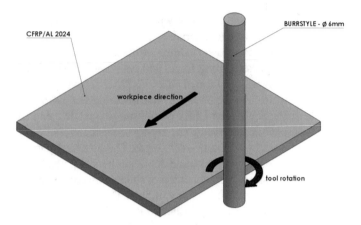

Fig. 1.3 Up trimming process

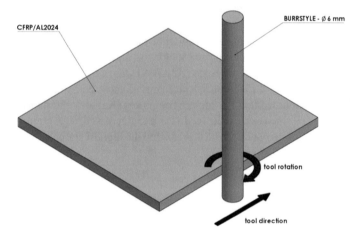

Fig. 1.4 Down trimming process

Haddad et al. [30, 31] only considered specimens that have similar roughness values as the abrasive water jet machining and ADS cutting plus a cutting condition where the surface roughness was very high (poor surface quality). The results showed that the defects generated during the trimming process with a cutting tool were fibers pullout and resin degradation. These defects were mainly located in the layers with the fibers oriented at $-45°$ and $90°$. On the other hand, Saleem et al. reported that when using abrasive water jet and abrasive diamond processes, the defects generated had the form of streaks and were not dependent on the fiber orientation.

Damages increase with the increase of cutting forces. Hence, Kalla et al. [32] proposed a neural network model to predict cutting forces during the trimming. This work utilized mechanistic modeling techniques for simulating the cutting of carbon fiber-reinforced polymers (CFRPs) with a helical end mill. A methodology was developed for predicting the cutting forces by transforming specific cutting energies from orthogonal cutting to oblique cutting. They gave acceptable results for the unidirectional specimen, but there were significant deviations observed with multilayer composites.

Haddad et al. [33] investigated the influence of tool geometry and cutting condition on surface defects and dust generated during the trimming. For conducting the studies on trimming, 20 layers of 5.2-mm-thick carbon fiber-reinforced plastic (CFRP) composite had been used. Machining experiments were conducted at a standard cutting speed (Feed speed (mm/min) = 500, 1000, and 1500 Cutting speed (m/min) = 150 and 250) and high-speed trimming (Feed speed (mm/min) = 125, 250, and 500, Cutting speed (m/min) = 350,700, and 1400) using full factorial design. The defects seemed to increase with the increase in feed speed or decrease in cutting speed in case of burr tool. However, the effects of the cutting parameters were completely different considering the four flute end mill.

Since the studies of trimming composite material are extremely limited, this book aims to expand the gap in the trimming of composite materials. Previous studies used CFRP laminates but vary in terms of stacking sequence. The project described in this book includes another element as a core material of the aluminum plate in the middle of CFRP laminates. A statistical analysis of variance (ANOVA) is employed to indicate the impact of cutting parameters on surface roughness. After each series of experiments, surface roughness tester is applied to check the roughness of machined surface.

1.4.5 Grinding

Grinding is one of the compulsory methods in fabricating products with composite materials and it is usually the final operation in the assembly of structural laminates. The study of Chockalingam et al. [34] aimed to develop an empirical model to predict the surface roughness of ground GFRP composite laminate with respect to the influence of grinding parameters. The significance of grinding parameters and the interaction effects of three factors on the grinding of GFRP composite were analyzed. An empirical equation was developed to attain minimum surface roughness in GFRP laminate grinding. They concluded that the most significant factors which influence the grinding surface roughness model are depth of cut, speed, and the interactions of speed, feed, and depth of cut. They observed that GFRP laminate grinding was very different from metal grinding and that its surface roughness did not depend solely on equivalent chip thickness alone. The combination of machining parameters played important roles in surface roughness quality.

Hu and Zhang investigated the grinding performance of epoxy matrix composites reinforced by unidirectional carbon fibers, using an alumina grinding wheel. They focused on the effect of fiber orientations and grinding depths on the grinding force and surface integrity and on the grinding mechanisms, with a comparison to orthogonal cutting. They found that greater grinding forces occurred at a fiber orientation between 60° and 90°, but poorer grinding surface finish took place between 120° and 180°. The surface integrity was highly dependent on the fiber orientation and the depth of grinding. The depth of the damage-affected zone increased with the increment of grinding depth for all the fiber orientations studied.

1.5 Laser Machining

LASER stands for light amplification by stimulated emission of radiation, which is the underlying principle for the generation of all types of lasers. Laser machining is a thermal process that removes material by melting and vaporization. The process parameters that influence laser cutting are power, traverse speed, assist gas pressure, and workpiece material composition and architecture. Laser machining offers several advantages over traditional machining processes due to no contact between the tool and the workpiece, and hence, there are no cutting forces, no tool wear, and no part distortion because of mechanical loading. One of the problems in laser cutting is material changes and strength reduction due to the formation of a heat-affected zone (HAZ), the formation of kerf tapers and a decrease in cutting efficiency as thickness of workpiece increases. Another problem in laser machining is the generation of hazardous chemical decomposition products. Laser cutting of aramid/epoxy produces large quantities of hydrogen cyanide, which may pose a considerable health risk [27].

Kwang-woon et al. [35] utilized the ultra-high-speed laser cutting on two kinds of CFRP with long or short carbon fibers using high brightness cw disk laser with a scanner head. The objectives of the research were to obtain better cut quality with submillimeter-sized HAZ and to confirm the cutting possibility for 3-mm-thick CFRP sheets. The effect of laser cutting parameters on CFRP cut qualities such as HAZ, kerf width, and kerf depth was evaluated. Kwang-woon et al. [35] concluded that the HAZ and kerf width at the cut surfaces and the cross sections became narrower and smaller with increasing the cutting speed. Furthermore, the increase of the laser power from 2 to 5 kW can reduce the processing time for full cutting, without degradation of good quality of narrow HAZ (less than 50 μm).

1.5.1 Abrasive Water Jet Machining

Abrasive water jet (AWJ) is one of the non-conventional machining methods of composite machining. Hashish [18] used AWJ in trimming of aircraft carbon fiber-reinforced plastic (CFRP) parts. He concluded that AWJ was very useful for trimming a wide angle of relatively large aerostructures. AWJ provides a good quality of machined surface but only limited shape can be machined and there is high cost in setting up as well as for its maintenance [36]. AWJ processing induces no thermal damage in the machined surface but the equipment is expensive. Furthermore, it is difficult to process as the jet does not penetrate. Sometimes, it is not suitable for the diverse application and wide varieties of CFRP parts [37].

1.5.2 Electrical Discharge Machining

Electrical discharge machining (EDM) is a non-traditional manufacturing technique that has been widely used in the production of tools and dies throughout the world in recent years. The most important performance measure in EDM is the surface roughness. Lodhi et al. [38] studied the effect and optimization of machining parameters on surface roughness in an EDM operation by using the Taguchi method. The study was conducted under varying gap voltage, discharge current, and pulse-on time. An orthogonal array, the signal-to-noise (S/N) ratio, was examined. The analysis of variance (ANOVA) was performed to the study the surface roughness of CFRP composite. They observed that the discharge current was the most influential factor on the surface roughness.

Habib [39] studied the electrical discharge machining (EDM) of carbon fiber-reinforced plastic (CFRP) material by developing an appropriate machining strategy for a maximum process criteria yield. A feed-forward back-propagation neural network was developed to model the machining process. The three most important parameters were considered, i.e., material removal rate, tool electrode wear rate, and surface roughness. The results showed more effective nature of neural networks in indicating the electrical discharge machining parameters. Well-trained neural network models provided fast, accurate, and consistent results, making them superior to all other techniques.

The machining of composites reviewed in this chapter is summarized in Fig. 1.5.

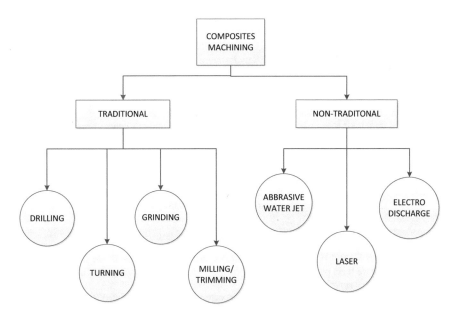

Fig. 1.5 Type of composites' machining

References

1. Bullen, N. (2010). Unified Structures. *Manufacturing Engineering Magazine.* Retrieved form http://www.sme.org/MEMagazine/Article.aspx?id=19179&taxid=1411.
2. Morey, B. (2009). Innovation drives composite production. *Manufacturing Engineering Magazine, Society of Manufacturing Engineer Editor, 142*(3), 49–60.
3. Morey, B. (2007). Composites challenge cutting tools. *Manufacturing Engineering Magazine Society of Manufacturing Engineer Editor, 138*(4), AT6–AT11.
4. Pop, P. A., Ungur, P., Lopez-Martinez, J., & Bejinaru-Mihoc, G. (2010, July). Manufacturing process and applications of composite materials. http://doi.org/10.15660/AUOFMTE.2010-2. 1896.
5. Rahim, E. A., Mohid, Z., Hamzah, M. R., Yusuf, A. F., & Rahman N A. (2014). Performance of tools design when helical milling on carbon fiber reinforced plastics (CFRP) aluminum (Al) Stack. *Applied Mechanics and Materials,* 465–466, 1075–1079. http://10.4028/www.scientific.net/AMM.465-466.1075.
6. Karpat, Y., Bahtiyar, O., & Değer, B. (2012). Milling force modelling of multidirectional carbon fiber reinforced polymer laminates. *Procedia CIRP, 1,* 460–465. https://doi.org/10.1016/j.pro cir.2012.04.082.
7. Andrei, G., Dima, D., & Andrei, L. (2006). Lightweight magnetic composites for aircraft applications. *Journal of Optoelectronics and Advanced Materials, 8*(2), 726–730.
8. Mangalgiri, P. D. (1999). Composite materials for aerospace applications. *Buletin of Materials Science, 22*(3), 657–664. https://doi.org/10.1007/BF02749982.
9. Liu, D., Tang, Y., & Cong, W. L. (2012). A review of mechanical drilling for composite laminates. *Composite Structures, 94*(4), 1265–1279. https://doi.org/10.1016/j.compstruct.201 1.11.024.
10. Sinmazçelik, T., Avcu, E., Özgür, M., & Çoban, O. (2011). A review : Fibre metal laminates, background, bonding types and applied test methods, *32,* 3671–3685. http://doi.org/10.1016/ j.matdes.2011.03.011.
11. Mazumdar, S. (2001). *Composites manufacturing: Materials, product, and process engineering.* Boca Raton, Florida: CRC Press.
12. Vogelesang, L. B., & Vlot, A. (2000). Development of fibre metal laminates for advanced aerospace structures. *Journal of Materials Processing Technology, 103*(1), 1–5. https://doi.or g/10.1016/S0924-0136(00)00411-8.
13. Davim, J. P., & Reis, P. (2005). Damage and dimensional precision on milling carbon fiber-reinforced plastics using design experiments. *Journal of Materials Processing Technology, 160,* 160–167. https://doi.org/10.1016/j.jmatprotec.2004.06.003.
14. Teti, R. (2002). Machining of composite materials. *CIRP Annals-Manufacturing Technology, 51*(2), 611–634. https://doi.org/10.1016/s0007-8506(07)61703-x.
15. Colligan, K., & Ramulu, M. (1992). Effect of edge trimming on composite surface plies. *Manufacturing Review, 5*(4), 274–283.
16. Wang, D. H., Ramulu, M., & Arola, D. (1995). Orthogonal cutting mechanism of graphite/epoxy composite. Part II: Multi-directional laminate. *International Journal Machine Tools Manufacture, 35*(12), 1639–1995. https://doi.org/10.1016/0890-6955(95)00014-0.
17. Abrate, S., & Walton, D. A. (1992). Machining of composite materials. Part I: Traditional methods. *Composites Manufacturing, 3*(2), 75–83. https://doi.org/10.1016/0956-7143(92)901 19-F.
18. Hashish, M. (2013). Trimming of CFRP aircraft components. *WJTA-IMCA Conference and Expo.* Retrieved from http://www.wjta.org/images/wjta/Proceedings/Papers/2013/A3%20-%2 0MH%20-%20Trimming.pdf.
19. Abrate, S., & Walton, D. A. (1992a). Machining of composite materials. Part II : Non-traditional methods. *Composites Manufacturing, 3*(2), 85–94. https://doi.org/10.1016/0956-7143(92)901 20-j.

20. Ozoegwu, C. G., Omenyi, S. N., Ofochebe, S. M., & Achebe, C. H. (2013). Comparing up and down milling modes of end-milling using temporal finite element. *Analysis, 3*(1), 1–11. https://doi.org/10.5923/j.am.20130301.01.

21. Gui, Y. L., Jian Feng, L., Jie, S., Wei Dong, L., & Liang Yu, S. (2010). A finite element model to simulate the cutting of carbon fiber reinforced composite materials. *Advanced Materials Research, 97–110*, 1745–1748. Retrieved from http://10.4028/www.scientific.net/AMR.97-101.1745.

22. Ozoegwu, C. G., Omenyi, S. N., Ofochebe, S. M., Obaseki, E., Uzoh, C. F., & Nwangwu, C. C. (2012). Comparison of up-milling and down-milling modes of end-milling process. *Research Journal in Engineering and Applied Sciences, 1*(2), 314–322.

23. Sadat, A. B. (2012). Machinability aspects of metal matrix composites (pp. 63–77). London : Springer-Verlag Limited. http://doi.org/10.1007/978-0-85729-938-3.

24. Krishnaraj, V., Zitoune, R., & Collombet, F. (2012). Study of drilling of multi-material (CFRP/Al) using Taguchi and statistical techniques. *Usak University Journal of Material Sciences, 2*, 95–109. Retrieved from http://dergipark.ulakbim.gov.tr/uujms/article/view/5000051 134/5000048355.

25. Zitoune, R., Krishnaraj, V., & Collombet, F. (2010). Study of drilling of composite material and aluminium stack. *Composite Structures, 92*(5), 1246–1255. https://doi.org/10.1016/j.com pstruct.2009.10.010.

26. Shyha, I., Soo, S. L., Aspinwall, D. K., Bradley, S., Dawson, S., & Pretorius, C. J. (2010). Drilling of Titanium/CFRP/Aluminium Stacks. *Key Engineering Materials, 448*, 624–633. https://doi.org/10.4028/www.scientific.net/KEM.447-448.624.

27. Sheikh-ahmad, J. Y. (2009). *Machining of polymer*. Abu Dhabi: Springer.

28. Palanikumar, K., Karunamoorthy, L., & Karthikeyan, R. (2004). Optimizing the machining parameters for minimum surface roughness in turning of GFRP composites using design of experiments, *20*(4), 373–378.

29. Kumar, S., Satsangi, P. S., & Sardana, H. K. (2012). Optimization of surface roughness in turning unidirectional glass fiber reinforced plastics (UD-GFRP) composites using polycrystalline diamond (PCD) cutting tool. *Indian Journal of Engineering & Materials Sciences, 19*, 163–174.

30. Haddad, M., Zitoune, R., Bougherara, H., Eyma, F., & Castanié, B. (2013). Study of trimming damages of CFRP structures in function of the machining processes and their impact on the mechanical behavior. *Composites: Part B, 57*, 136–143. https://doi.org/10.1016/j.composites b.2013.09.051.

31. Haddad, M., Zitoune, R., Eyma, F., & Castanie, B. (2013). Machinability and surface quality during high speed trimming of multi directional CFRP. *International Journal of Machining and Machinability of Materials, 13*(2/3), 289–310. https://doi.org/10.1504/ijmmm.2013.053229.

32. Kalla, D., Sheikh-ahmad, J., & Twomey, J. (2010). Prediction of cutting forces in helical end milling fiber reinforced polymers. *International Journal of Machine Tools and Manufacture, 50*(10), 882–891. https://doi.org/10.1016/j.ijmachtools.2010.06.005.

33. Haddad, M., Zitoune, R., Eyma, F., & Castanie, B. (2014). Study of the surface defects and dust generated during trimming of CFRP: Influence of tool geometry, machining parameters and cutting speed range. *Composites: Part A, 66*, 142–154. https://doi.org/10.1016/j.compos itesa.2014.07.005.

34. Chockalingam, P., Kok, C. K., & Vijayaram, T. R. (2013). Surface roughness prediction model for grinding of composite laminate using factorial design. *International Journal of Mechanical, Industrial Science and Engineering, 7*(4), 166–170.

35. Kwang-woon, J., Yousuke, K., & Seiji, K. (2013). Ultra high speed laser cutting of CFRP using a scanner head, *42*(2), 9–14. Retrieved from http://www.jwri.osaka-u.ac.jp/publication/ trans-jwri/pdf/422-03.pdf.

36. Arisawa, H., Akama, S., & Niitani, H. (2012). High-performance cutting and grinding technology for CFRP (Carbon Fiber Reinforced Plastic). *Mitsubishi Heavy Industries Technical Review, 49*(3), 3–9. Retrieved from http://www.mhi.co.jp/technology/review/pdf/e493/e4930 03.pdf.

37. Yashiro, T., Ogawa, T., & Sasahara, H. (2013). Temperature measurement of cutting tool and machined surface layer in milling of CFRP. *International Journal of Machine Tools and Manufacture, 70,* 63–69. https://doi.org/10.1016/j.ijmachtools.2013.03.009.
38. Lodhi, B. K., Verma, D., & Shukla, R. (2014). Optimization of machining parameters in EDM of CFRP composite using TAGUCHI technique. *International of Mechanical Engineering and Technology (IJMET), 5*(10), 70–77.
39. Habib, S. S. (2014). Modeling of electrical discharge machining of CFRP material through artificial neural network technique. Journal of Machinery Manufacturing and Automation, *3*(1), 22–31. Retrieve from http://www.academicpub.org/jmma/paperInfo.aspx?paperid=15310.

Chapter 2
Cutting Parameters and the Machinability Performance

2.1 Types of Cutting Parameters

Optimizing cutting parameters is very significant to obtain good machined surface and meet engineering specifications. It is also can save energy, reduce waste, save processing time, and increase tool life [1]. Generally, there are four types of cutting parameters normally associated with machining operation, i.e., cutting speed, spindle speed, depth of cut, and feed rate [2–6]. All of these parameters have been identified as the influential factors in determining the surface quality of every machined part.

Most of the researchers focused on four cutting parameters during their studies on optimization in composite machining. They are spindle speed, cutting speed, depth of cut, and feed rate [7–9]. In general, the best machined surface quality is being determined by the kind of material being cut, and the size and type of the cutter used, width and depth of cut, method of application, and speed available are factors relating to machinability performance.

2.1.1 Cutting Speed (m/min)

The cutting speed expressed in meters per minute (m/min) must not be confused with the spindle speed which is expressed in revolution per minute (rpm). Cutting speed represents the rate of the cutter passed over the surface of the machined part, whereby the spindle speed is obtained by calculating from a selected cutting speed.

The cutting speed of a metal may be defined as the speed, in surface feet per minute or linear feet per minute (sf/min or mm/min) that a given tooth (flute) at which the metal may be machined efficiently. When the work is machined on a milling machine, the cutter must be revolved at a specific number of (r/min), depending on its diameter to achieve the proper cutting speed. In workshop practices, the machinist used spindle speed, because the machine only reads the value of the spindle rather than cutting speed. Further, cutting speed can be calculated using the expression below.

$$V_c = \frac{\pi * D * N}{1000} \tag{2.1}$$

where D is the diameter of the cutter, and N is revolution per minute (spindle speed).

The cutting speed is given as a set of constants that are normally available from the cutting tools' manufacturer but will always be subject to adjustment depending on the cutting condition.

2.1.2   Spindle Speed (RPM)

The spindle speed is the rotational frequency of the machine spindle which is measured in revolution per minute (rpm). The preferred spindle speed is determined by calculating from the desired cutting speed and incorporating the diameter of cutter using the following formula:

$$N = \frac{V_c * 1000}{\pi * D} \tag{2.2}$$

where N is a spindle speed, V_c is cutting speed, and D is the diameter of the cutter.

Selecting the correct spindle speed to use when cutting any material with a rotating tool has always been a challenge. This is particularly true of materials like composite where there is such variation in cutting properties.

2.1.3   Feed Rates (mm/min)

Feed is the rate at which advancement of the tool toward the part and measured in feet, inches, or millimeters per time period. It is also known as the table feed, machine feed, or feed. The equation for the feed rate for the process is as follows (Table 2.1):

$$\text{Feed rate} = f * \text{CPT} * N \tag{2.3}$$

where f is the number of teeth (flutes) in milling cutter, CPT is the chip per tooth for a particular cutter, and N is the spindle speed of the cutter.

Table 2.1 Research on trimming of CFRP

Researchers, Year	Control variables	Process	Findings
Slamani et al. [29]	Tool: Polycrystalline diamond (PCD) Cutting parameters: cutting speed, feed rate, and fiber orientation	Comparison of surface roughness using high-speed CNC and high-speed robotic trimming of CFRP laminate	Significant relationship between surface quality and ply orientation whatever machining process and cutting conditions used $-45°$ ply orientation, effect on feed rate constituted the most significant effect on surface roughness (CNC = 61.20% and robotic = 38.47%) and followed by cutting speed (CNC = 23.94% and robotic = 33.72%) $+45°$ ply orientation, effect of cutting speed constituted the most significant effect on CNC process, 63.94% Optimum cutting condition for good surface roughness: Robotic trimming: low feed, low cutting speed; CNC trimming: medium feed and high cutting speed
Mohamed et al. [8]	Tool: Polycrystalline diamond (PCD) Cutting parameters: spindle speed, feed rate, and depth of cut	Edge trimming via up milling of CFRP/Al2014 DoE: Two-level full factorial design	Most significant parameters on surface roughness: depth of cut and feed rate Surface roughness of CFRP was 1.778 μm and Al2024 was 0.938 μm at the setting of 1860 rpm spindle speed, 620 mm/min feed rate, and 0.12 mm depth of cut
Haddad et al. [14]	Tool geometry, cutting speed, feed rate, and cutting distance	High-speed trimming of CFRP	Quality of machined surface mainly affected by cutting speed and cutting distance

2.1.4   Depth of Cut (DOC)

The measurement normally in inches or millimeters of how wide and deep the tool cuts into the workpiece. Argawal defined the DOC as the thickness of the metal that is removed in one cut. It is the perpendicular distance measured between the machined surface and non-machined surface on the workpiece.

2.2   Surface Roughness

Surface roughness is a characteristic that could influence the dimensional precision, the performance of mechanical pieces, and production costs [10]. Quality of surface roughness is one of the most important requirements in machining operations. Maintaining good surface quality usually involves additional manufacturing cost and loss of productivity; therefore, the optimization of cutting parameters to minimize surface roughness needs to be undertaken [11, 12].

Arithmetic mean value (Ra) is used to describe surface roughness because it is well recognized and widely used for international parameter of roughness. It is the arithmetic mean of the actual departures of roughness profile from the mean line. Ra is reported in microns (1×10^{-6}m $= 0.001$ mm). The theoretical expression for surface roughness can be seen as follows: Ra is defined as

$$Ra = (a + b + c + d + \cdots)/n \tag{2.4}$$

The actual values are presented by the ordinates a, b, c, d, and the number of readings is presented by n. μm (micrometer, or micron) is the unit that generally used for the surface roughness.

Davim and Reis [10] carried out statistical study on surface roughness in milling of CFRP using two cemented carbide end mill. The variables considered here were cutting speed and feed rate and damage in milling laminates plates. They discovered for both end mills, and it was possible to obtain surface between 1 and 3 μm of surface roughness according to the cutting parameters used. Their observation revealed that feed rate presents statistical significance on surface roughness compared to other factors.

Kumar et al. [13] studied the optimization of surface roughness in turning unidirectional glass fiber-reinforced plastics (UD-GFRPs) using PCD tool. Taguchi method was chosen to carry out this experiment. The input parameters taken up for optimization were feed rate, cutting speed, and depth of cut along with tool nose radius, tool rake angle, and cutting environment. The parameters that had higher influence are feed rate and cutting speed while the cutting environment did not influence the surface roughness significantly.

Research done by Mohamed et al. [8] studied on machining parameters optimization of trimming operation (up milling) using two-level full factorial design. They considered the machining parameters, i.e., spindle speed, feed rate, and depth of cut. The aim was to analyze the influence factors and the interaction of these parameters with respect to surface quality produced. The analysis of variance (ANOVA) showed that depth of cut and feed rate were the most significant parameters to the overall performance of surface quality. Surface roughness of CFRP was found to be 1.778 μm, and Al2024 was found to be 0.983 μm.

2.3 Machined Surface Quality of Composites

There are various methods of evaluating machined surfaces and each of them has its own unique characteristics corresponding to the quality requirements. The choice of evaluation technique depends on the equipment availability, the evaluator, and the established terms in the quality control guidelines. Residual stress induced and surface topography are the indicators for mechanical performance of homogeneous material. Since the inhomogeneous material like CFRP does not develop residual stress in machining, the quality of the machined surface is evaluated based on surface profilometry and visual techniques.

Quality of a machined surface for composites is dependent on fiber orientation and type of fiber used [14]. From the literature review conducted, it was seen that at fiber orientation of 300° and 450° the surface was very poor, since the fiber was cut by combined compressive and bending stress, while better surface was obtained at fiber orientation of 90°. Moreover, other factors that influence surface roughness are tool wear, feed rate, and cutting speed. Studies show that surface roughness increases with increase in tool wear, feed rate, and decreasing in spindle speed.

Generally, there are two approaches in characterizing a machined surface. First is roughness parameter (Ra, Rq, Rz, Rt). Second approach is statistical parameters such as skewness, kurtosis, and frequency height distributions. The various roughness parameters are arithmetic average height (Ra), root mean square height (Rq), peak-to-valley height (Rt), valley-to-mean height (Rv), and ten-point average height (Rz). However, studies have shown that roughness parameters Ra and Rq were the most common approach characterizing machined with respect to fiber orientation [14] (Agawal 2012; Azmi 2012). So, the preferred roughness parameter for representing the surface features of composites is arithmetic average height, Ra. The role of these roughness parameters is to evaluate the surface produced by a machining process and to optimize cutting process parameters such as cutting speed, feed rate, and depth of cut. It has been shown that if the value of surface roughness is lower, the quality of machined surface is better. Roughness values also indicate changes in the mechanical properties of machined FRP. Studies have shown that with increasing roughness the fatigue strength and impact strength decreased.

Roughness is measured on a machined surface by using stylus profilometer, which is the common instrument used in the industry. Roughness value is given by the vertical displacement of a diamond stylus tip, which moves along the machined surface. Result of roughness measurement greatly depends on the stylus path, since fiber direction changes from layer to layer. Neebu suggested two methods either taking roughness measurements by keeping the stylus in one layer and take readings at different locations of this layer or take readings at different location for different layers and take the average. Things that need to be carefully considered during measurement process such as matrix smearing, fiber protruding, and fiber clinging to the stylus tip will obliterate the reading and will not give an exact description of the surface. So, a visual inspection in combination with the profilometer reading is necessary to quantify surface topography. However, due to the equipment availability only Mitutoyo SJ-301 was used for surface roughness tester to record the value of machined surface. Since measuring both materials at the same time would cause error reading to the surface roughness tester and accurate data cannot be obtained, the author decided to measure the surface roughness separately as suggested by Mohamed et al. [8].

2.4   Design of Experiments (DoE)

The design of experiments (DoE) allows us to optimize processes from a reliability viewpoint and product costs. In a domain such as machining, the use of the DoE method can be of great help in terms of reliability and product cost [15].

DoE is a systematic approach to understand the process and product parameters that affects the response variables. It is defined as a series of tests in which purposeful changes are made to input factors so that the causes for significant changes in the output responses can be identified. It is a mathematical tool that generates, summarizes, and evaluates to ensure their feasibility. By doing this, time spent in conducting experiments can be minimized and the quality of experiments can be controlled. In short, DoE approaches are able to minimize trails and save a lot of time and cost [16]. There are a lot of DoE software available in the market such as Design Experts, Mini Tab, and statgraphics. DoE is an integrated software which is a set of mathematical and statistical techniques useful for modeling and analyzing of complex process optimization [17]. There are few types of DoE, i.e., full factorial designs, fractional factorial designs, Placket Burman designs, response surface designs, Taguchi designs, and mixture designs.

The optimization of turning parameters on surface roughness of glass fiber-reinforced plastic was carried out by Palanikumar et al. [18]. By using DoE and analysis of variance (ANOVA), the authors concluded that the technique was so convenient and economical to predict optimal cutting parameters.

Benyounis and Olabi [19] stated that Taguchi method was one of the powerful optimization techniques to improve product quality and reliability at low cost. Based on the research done by Sait et al. [20] and Tsao and Hocheng [21], computational time done by Taguchi design method was at the medium as compared to response surface methodology (RSM) and factorial design.

Rajmohan et al. [11] performed the investigation on optimal design of cutting parameters for drilling hybrid matrix composite. The effect of input parameters namely spindle speed, feed rate, and weight percentage on the thrust force and surface roughness was studied in this experiment. They applied RSM and central composite design (CCD) for modeling, optimization, and analysis. This investigation proved that the proposed approach could be useful to improve the performance of the process.

Desai and Rana [22] studied the optimum drilling parameters, spindle speed, and feed rate on CFRP laminates to get optimum cutting conditions. DoE methodology by full factorial design was used in multiple objective optimizations (using software, Mini Tab 16) to find the optimum cutting conditions for defect free drilling.

Haddad et al. [23, 24] conducted high-speed trimming of a multidirectional CFRP using unused and used burrs' tools to investigate the influence of the machining parameters (feed speed, cutting speed, and cutting distance) on the cutting forces, machining temperature, and the machined surface quality. The experiment was conducted using full factorial design and the ANOVA was used to analyze the experimental results. They found that the machining parameters had a significant influence on the variation of the machined surface quality and the cutting forces.

The application of Taguchi methods in optimizing the cutting parameters (depth of cut, cutting speed, and feed rate) of end milling process under dry condition was studied by Pang et al. [25]. The surface roughness of the machined composite and the cutting forced was measured. Taguchi method was employed to determine the best combination of cutting parameters could provide the optimal machining response conditions, i.e., the lowest surface roughness and the lowest cutting force values.

Mohamed et al. [8] utilized DoE, two-level full factorial design to investigate machining parameters optimization for trimming operation of carbon fiber-reinforced plastic laminated with aluminum grade 2024 (CFRP/Al2024) using milling machine. They analyzed the influence factors and the interaction between these cutting parameters and determined the optimum machining parameters for minimizing surface roughness. Surface roughness of CFRP is found to be 1.778 μm and Al2024 is found to be 0.983 μm at the setting of spindle speed 1860 rpm, feed rate 620 mm/min, and depth of cut 0.12 mm, respectively.

2.4.1   Two-Level Factorial Designs

The traditional approach to experimental planning, i.e., one-at-a-time designs, involves a lot of time and effort and does not consider all of the interaction factors. The most efficient way for enhancing the value of research and to reduce time in process development is through experimental designs. Thus, from a few but well-conducted number of experiments, a picture of response surface can be achieved [17].

Factorial designs facilitate the simultaneous study of the effects that several factors may have on the optimization of a certain process. This method determines which factors have the important effects on the response and how the effect of one factor varies with the level of the other factors. The effects are the differential quantities showing how a response changes as the levels of one or more factors are changed. Furthermore, these factorial designs allow measuring the interaction between each different group of factors. In many optimizations of the processes, the interactions play significant roles. Without utilizing factorial experiments, some important inter-actions may remain unconsidered and the overall optimization may not be achieved. One of the simplest types of factorial designs used in experimental work is one having two levels ($2k$). In a $2k$ factorial design experiment, each factor may be assigned two levels: low (-1) and high ($+1$). If k factors are considered, then $2k$ measurements are required to perform a factorial design analysis [26].

Mohamed et al. [7] studied the effect of cutting parameters on the surface structure of hybrid composite CFRP/AL2024. Two-level factorial design was the chosen DoE methodology in this research. The research aimed to study the interaction effects and significant factors of cutting parameters on the surface quality and optimize the cutting parameter for the surface quality of CFRP/Al2024 1–2 μm. The trimming process test was performed under dry conditions using Kennametal burr tool of burr tools of 6 mm in diameter. The factors investigated were spindle speed, feed rate, and depth of cut; meanwhile, profile roughness parameters (Ra) of CFRP and Al2024 were the response variables.

The research that investigated the effects of the most significant parameters such as spindle speed, depth of cut, feed rate, and tool size on the surface roughness was done by Noorani et al. [27]. The classical method of design of experiments (DoE) was chosen for the research. A two-level, four-factor full factorial experiment was used to select the best combination of factors level that would minimize the surface roughness. A statistically designed experiment was used to determine the processing factors that affected the surface roughness of the aluminum samples made by the CNC milling machine. The significant factors, their interactions, optimum setting, and the physical interpretation of the effects of the process parameters on surface roughness were presented in this research.

2.4.2 The 2^3 Factorial Design

The two-level full factorial design is utilized to determine which factors contribute to an important effect on the output of the reaction and distinguishing the relations between factors.

$2k$ runs with each factor at two levels are required for factorial design. The number of factors is represented by k while the number of levels is represented by 2. There are two levels for each factor which is known as low and high (often labeled + and −). Cox and Reid [28] have tested all possible factor-level combinations and found that two-level full factorial design is the simplest form of factorial designs. In this experiment, eight runs of experiment are required for a two-level full factorial design of three factors. To increase the accuracy of the value on the effect and gain other information on the process variation, the run was duplicated with a new batch of work material. The experiments are repeated and run in different times to escalate

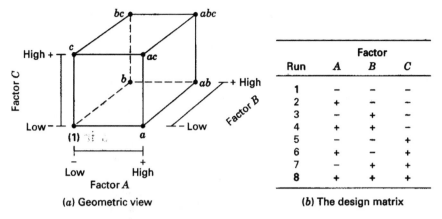

	Factor		
Run	A	B	C
1	−	−	−
2	+	−	−
3	−	+	−
4	+	+	−
5	−	−	+
6	+	−	+
7	−	+	+
8	+	+	+

(a) Geometric view   (b) The design matrix

Fig. 2.1 2^3 factorial design [28]

the accuracy of model. Two-level full factorial design is a very practical design for introductory analysis and able to put off unimportant factors and also allows initial study of interactions. Thus, it gives a full understanding of the structure of the design. A, B, and C are the three factors of 2^3 design which each of them has two levels. It can be understood by referring to Fig. 2.1. Three factors and two levels are commonly done as they are attainable and manageable most of the times. The geometric view of design matrix shows orthogonality all possible corner experiments run.

2.5 Research on Machining of Fiber Metal Laminates

Machining of FML depends on the fibers orientation, type of binder used, and metal sheet which will affect the selection of machining parameters. This is important in order to avoid the insufficient surface finish with high material removal rate (MMR) which leads to the reduction in manufacturing costs.

Slamani et al. [29] conducted a study on trimming of CFRP laminates using high-speed CNC versus high-speed robotic system. The controlled factors were cutting speed, feed rate and fiber orientation, whereas the machined surface quality was the response factor. They found that a significant relationship between surface quality and fiber orientation arrangement. For −45° ply orientation, the effect of the feed rate constituted the most significant factor whatever the machining process with contributions of 61.2028 and 38.4748% for the CNC trimming and robotic trimming process, respectively. The next largest factor is the cutting speed with contributions of 23.9361 and 33.7195% for the CNC trimming and robotic trimming process, respectively. The cutting speed with a contribution of 63.9456% is the largest factor which influences the average surface roughness of a +45° ply orientation.

A study on trimming CFRP stacked with aluminum grade 2024 was done by Mohamed et al. [8]. Three cutting parameters were analyzed namely spindle speed, feed rate, and depth of cut with the surface roughness as a response variable. The trimming process was performed on CNC milling machine using polycrystalline diamond (PCD) end mill cutter of 6 mm diameter via up milling. Two-level full factorial experimental design was employed to plan systematic experimental work and performing analysis. They found that depth of cut and feed rate were the most significant parameters to the overall performance of surface roughness quality. Surface roughness of CFRP was 1.778 μm, and Al2024 was 0.983 μm at the setting of 1860 rpm spindle speed, 620 mm/min feed rate, and 0.12 mm depth of cut, respectively.

Trimming damages of CFRP structure in function of various machining process (burr tool machining, abrasive water jet machining, and abrasive diamond cutter machining) and their impact on the mechanical behavior via down milling approaches were investigated by Haddad et al. [23, 24]. The controlled factor was tool geometry, cutting speed, feed rate, and cutting distance. Response variable was the surface roughness. They observed that the defects predominantly occurred due to cutting conditions and tool geometry. During high cutting conditions, the quality of the machined surface was affected by cutting speed followed by cutting distance and finally by feed speed. They also found that feed speed was the major parameter affecting surface roughness under standard cutting conditions.

Haddad et al. [14] investigated the influence of tool geometry and cutting conditions on machined surface quality and the dust generated during trimming. This study considered tool geometry, cutting speed, feed rate, and cutting distance as controlled variables. Analysis of variance (ANOVA) was employed to estimate the effect of the controlled factors on the cutting forces and surface roughness. They found the quality of the machined surface was mainly affected by the cutting speed (with a contribution percentage of 49.1%) followed by cutting distance (with a contribution percentage of 18.67%).

Following a detailed literature review, it has emerged that determining of machining parameters and response variables has become important criteria in composite machining. However, the effectiveness of machining process itself in terms of manufacturing cost reduction has become a major challenge. By considering the various machining parameters as proposed by the most of the researchers, it can be initially concluded that control factors of spindle speed, feed rate, and depth of cut as well as tool geometry have greatly affected the machinability and machined surface performance of composite materials.

All of the composites machining described previously either traditional or nontraditional employ optimization approach which having cutting parameters as the control factors and surface finished as a response variable. The method is a straightforward experiment involving the manipulating of quantitative data, to generate statistically analyzable data which giving a clear and unambiguous picture of the work under consideration. Noted ably, three types of DoE employed for traditional composite machining namely Taguchi method, response surface methodology, and two-level full factorial design. However, the choice of design methodology generally depends on the experimental objective and the number of controlled factors.

The selection of DoE methods mentioned above depends on the preciseness of the objective and on the number of factors to be investigated. The precision of the objective can be translated in terms of complexity of the model that will be used to analyze the results of the experiments. The level of complexity can be divided into three categories. First, linear model studies the main effect of the individual factor only. Second, linear model of interaction which takes into account the way one factor modifies the effect on another factor. The third is quadratic models which include square terms in addition to linear terms and interactions having model response surface with curvature. Therefore, this study will start up with linear model to investigate the main effect of each factor, and if the curvature detected, a quadratic model will be used. The best approach for this research is to use a two-level full factorial design. If the curvature detected, the extended full factorial design will be applied of having quadratic model.

References

1. Pang, J. S., Ansari, M. N. M., Zaroog, O. S., Ali, M. H., & Sapuan, S. M. (2013). Taguchi design optimization of machining parameters on the CNC end milling process of halloysite nanotube with aluminium reinforced epoxy matrix (HNT/Al/Ep) hybrid composite. *HBRC Journal, 10*(2), 138–144. https://doi.org/10.1016/j.hbrcj.2013.09.007.
2. Yang, Y., Chuang, M., & Lin, S. (2009). Optimization of dry machining parameters for high-purity graphite in end milling process via design of experiments methods. *Journal of Materials Processing Technology, 209,* 4395–4400. https://doi.org/10.1016/j.jmatprotec.2008.11.021.
3. Alberti, M., Ciurana, J., & Casadesus, M. (2005). A system for optimising cutting parameters when planning milling operations in high-speed machining. *Journal of Materials Processing Technology, 168*(1), 25–35. https://doi.org/10.1016/j.jmatprotec.2004.09.092.
4. Aggarwal, A., & Singh, H. (2005). Optimization of machining techniques—A retrospective and literature review. *Sadhana, 30*(6), 699–711. https://doi.org/10.1007/BF02716704.
5. Vivancos, J., Luis, C. J., Costa, L., & Ort, J. A. (2004). Optimal machining parameters selection in high speed milling of hardened steels for injection moulds. *Journal of Material Processing Technology, 156,* 1505–1512. https://doi.org/10.1016/j.jmatprotec.2004.04.260.
6. Dharan, C. K. H., & Won, M. S. (2000). Machining parameters for an intelligent machining system for composite laminates. *International Journal of Machine Tools and Manufacture, 40*(3), 415–426. https://doi.org/10.1016/S0890-6955(99)00065-6.
7. Mohamed, S. B., Mohamad, W. N. F., Muhamad, M., Ismail, J., Yew, B. S., Mohd, A., et al. (2016). The effects of cutting parameters on surface texture of hybrid composite CFRP/AL2024. *Materials Science Forum, 863,* 111–115. https://doi.org/10.4028/www.scientific.net/MSF.863.111.
8. Mohamed, S. B., Wan Mohamad, W. N. F., Seok, Y. B., Minhat, M., Kasim, M. S., Ibrahim, Z., et al. (2015). Machining parameters optimization for trimming operation in a milling machine using two level factorial design. *Applied Mechanics and Materials, 790,* 105–110. https://doi.org/10.4028/www.scientific.net/AMM.789-790.105.
9. Venkata, K., Krishnam, M., & Janardhana, G. R. (2011). Optimization of cutting conditions for surface roughness in CNC end milling. *International Journal of Precission Engineering and Manufacturing, 12*(3), 383–391. https://doi.org/10.1007/s12541-011-0050-7.
10. Davim, J. P., & Reis, P. (2005). Damage and dimensional precision on milling carbon fiber-reinforced plastics using design experiments. *Journal of Materials Processing Technology, 160,* 160–167. https://doi.org/10.1016/j.jmatprotec.2004.06.003.

11. Rajmohan, T., Palanikumar, K., & Kathirvel, M. (2012). Optimization of machining parameters in drilling hybrid aluminium metal matrix composites. *Transactions of Nonferrous Metals Society of China, 22*(6), 1286–1297. https://doi.org/10.1016/S1003-6326(11)61317-4.
12. Sait, A. N., Aravindan, S., & Haq, A. N. (2008). Optimisation of machining parameters of glass-fibre-reinforced plastic (GFRP) pipes by desirability function analysis using Taguchi technique. *The International Journal of Advanced Manufacturing Technology, 43*(5–6), 581–589. https://doi.org/10.1007/s00170-008-1731-y.
13. Kumar, S., Satsangi, P. S., & Sardana, H. K. (2012). Optimization of surface roughness in turning unidirectional glass fiber reinforced plastics (UD-GFRP) composites using polycrystalline diamond (PCD) cutting tool. *Indian Journal of Engineering & Materials Sciences, 19*, 163–174.
14. Haddad, M., Zitoune, R., Eyma, F., & Castanie, B. (2014). Study of the surface defects and dust generated during trimming of CFRP: Influence of tool geometry, machining parameters and cutting speed range. *Composites: Part A, 66*, 142–154. https://doi.org/10.1016/j.composit esa.2014.07.005.
15. Zenia, S., Ayed, L. B., Nouari, M., & Delamézière, A. (2014). Numerical analysis of the interaction between the cutting forces, induced cutting damage, and machining parameters of CFRP composites. *International Advanced Manufacturing Technology*. https://doi.org/10.100 7/s00170-014-6600-2.
16. Anderson, M. (1997). Design of Experiments. *American Institute of Physics*, 24–26. Retrieved from https://www.researchgate.net/profile/Mark_Anderson43/publication/2280180 75_Design_of_Experiments/links/54997a390cf21eb3df60d33a.pdf.
17. Gonzales, A. G. (1998). Two level factorial experimental designs based on multiple linear regression models A tutorial digest illustrated by case studies. *Analytica Chimica Acta, 360*(1–3), 227–241. https://doi.org/10.1016/S0003-2670(97)00701-0.
18. Palanikumar, K., Karunamoorthy, L., & Karthikeyan, R. (2004). *Optimizing the machining parameters for minimum surface roughness in turning of GFRP composites using design of experiments, 20*(4), 373–378.
19. Benyounis, K. Y., & Olabi, A. G. (2008). Optimization of different welding processes using statistical and numerical approaches—A reference guide. *Advances in Engineering Software, 39*(6), 483–496. https://doi.org/10.1016/j.advengsoft.2007.03.012.
20. Sait, A. N., Aravindan, S., & Haq, A. N. (2009). Optimisation of machining parameters of glass-fibre-reinforced plastic (GFRP) pipes by desirability function analysis using Taguchi technique, *International Journal Advance of Manufacturing Technology (IJAMT)*, 581–589. http://dx.doi.org/10.1007/s00170-008-1731-y.
21. Tsao, C. C., & Hocheng, H. (2004). Taguchi analysis of delamination associated with various drill bits in drilling of composite material. *International Journal of Machine Tools and Manufacture (IJMACT), 44*, 1085–1090. https://doi.org/10.1016/j.ijmachtools.2004.02.019.
22. Desai, B., & Rana, J. (2012). Machining characterization of CFRP laminates with respect to drilling operation—A review. *Journal of Engineering Research and Studies, III*(IV), 8–12. Retrieved from http://www.technicaljournalsonline.com/jers/VOL%20III/JERS%20VOL%2 0III%20ISSUE%20IV%20OCTOBER%20DECEMBER%202012/Article%203%20Vol%20 III%20Issue%20IV.pdf.
23. Haddad, M., Zitoune, R., Bougherara, H., Eyma, F., & Castanié, B. (2013). Study of trimming damages of CFRP structures in function of the machining processes and their impact on the mechanical behavior. *Composites: Part B, 57*, 136–143. https://doi.org/10.1016/j.composites b.2013.09.051.
24. Haddad, M., Zitoune, R., Eyma, F., & Castanie, B. (2013). Machinability and surface quality during high speed trimming of multi directional CFRP. *International Journal of Machining and Machinability of Materials, 13*(2/3), 289–310. https://doi.org/10.1504/ijmmm.2013.053229.
25. Pang, J. S., Ansari, M. N. M., Zaroog, O. S., Ali, M. H., & Sapuan, S. M. (2014). Taguchi design optimization of machining parameters on the CNC end milling process of halloysite nanotube with aluminium reinforced epoxy matrix (HNT/ Al/ Ep) hybrid composite. *HBRC Journal, 10*(2), 138–144. https://doi.org/10.1016/j.hbrcj.2013.09.007.

26. Yahiaoui, I. (2010). Experimental design for copper cementation process in fixed bed reactor using two-level factorial design. *Arabian Journal of Chemistry, 3*(3), 187–190. https://doi.or g/10.1016/j.arabjc.2010.04.009.
27. Noorani, R., Farooque, Y., & Ioi, T. (2010). Improving surface roughness of CNC Milling machined aluminum samples due to process parameter variation, pp. 1–7. Retreived from http://citeseerx.ist.psu.edu/viewdoc/download?doi=10.1.1.624.3481&rep=rep1&type=pdf.
28. Cox, D. R., & Reid, N. (2000). *The theory of the design of experiments*. New York: Chapman & Hall/CRC.
29. Slamani, M., Gauthier, S., & Chatelain, J. (2016). Comparison of surface roughness quality obtained by high speed CNC trimming and high speed robotic trimming for CFRP laminate. *Robotics and Computer Integrated Manufacturing, 42,* 63–72. https://doi.org/10.1016/j.rcim. 2016.05.004.

Chapter 3
The Carbon Fiber-Reinforced Plastic Project

3.1 Overview of the Project

Activities in a research project may be structured into a series of tasks. Each task should be clearly defined and understood to ensure the possibility of introducing effective strategies and concepts for experimental design. Figure 3.1, on the next page, shows the main tasks in a standard experimental design which consist of sample preparations, experimental setup, data collection and analysis, and process validation. Besides analyzing the cutting parameters, the research project also optimizes the cutting parameter and calculates the percentage error between the predicted and actual value of surface roughness generated.

3.1.1 The Experimentation

Experiments were carried out based on the sequence table experimental design method determined by Design-Expert software. There were three parameters used, namely spindle speed, feed rate, and depth of cut. Effects of changes in all three parameters were reflected to the machined surface quality of the tested workpiece.

3.1.2 Workpiece Details

The carbon fibers impregnated with resin carbon prepreg of 16 layers and 4.0 mm thicknesses were used in this project. The materials which were manufactured by Cytec Aerospace Material were supplied by Composite Testing Laboratory Asia Sdn. Bhd (CTLA). The specimen used in the project is shown in Fig. 3.2.

© The Author(s), under exclusive license to Springer Nature Singapore Pte Ltd 2019   29
S. B. Mohamed et al., *Down Milling Trimming Process Optimization for Carbon Fiber-Reinforced Plastic*, SpringerBriefs in Applied Sciences and Technology, https://doi.org/10.1007/978-981-13-1804-7_3

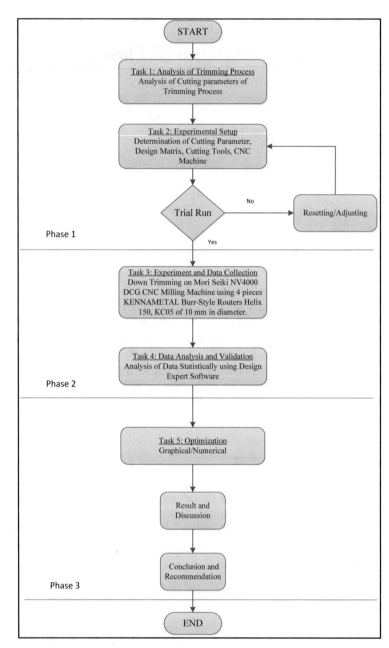

Fig. 3.1 Schematic illustration of research work

Fig. 3.2 CFRP/Al2024 specimen

3.1.3 Solid Carbide End Mill Cutting Tool

Carbide cutting tool is a type of cutting device used for metal cutting for an extensive variety of materials on CNC machines. The tool has high wear resistance, and it is efficient in a long production run. It also retains the cutting-edge hardness at high machining temperatures generated by high cutting speeds besides attaining to a superior surface quality. Carbide tools are less costly compared to polycrystalline coated diamond (PCD) tools [1].

In the case of glass and carbon fiber reinforcement, a good machined surface with a precise accuracy is required in machining of FRP parts [2]. Different types of material can be machined using carbide tools such as aluminum alloys and non-ferrous material such as plastics and composites [3, 4]. Generally, the diamond-based tool and carbide tool are used when a good machined surface with a precise accuracy is required in machining process. Despite the good performance of these two types of tool, there are still limitations. According to Sandvik Coromant technical data sheet (Arnett & Jacques, n.d.), carbide is strong but it tends to wear rapidly, and while PCD tool is very good in terms of wear resistance, they are brittle. Diamond and carbide tools are highly recommended for machining composite materials.

Initially, we proposed to apply a diamond cutter during the machining process, as PCD has been the preferred solution to conduct machining on aircraft stacking materials. After some slot of discussion between the CTRM's and CTLA's machinists and engineers, we decided to use solid carbide as this type of end mill can provide good results. Two pieces of Kennametal Burr-Style Routers Helix 15°, KCN05 of 6 mm in diameter, micrograin size of 5 μm, and the titanium aluminum nitride (TiAlN) coating were used, as shown in Fig. 3.3.

Table 3.1 shows the range of cutting condition suggested by the manufacturer for the tool to be used for machining purposes.

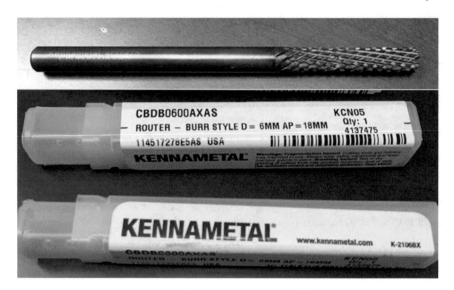

Fig. 3.3 Kennametal Burr-Style Routers Helix 15°, KCN05 of 6 mm in diameter

Table 3.1 Range of cutting of Kennametal Burr-Style Routers Helix 15°, KCN05 of 6 mm in diameter

Description	Units	Min	Max
Cutting speed	m/min	100	150
Feed per tooth	mm/th	0.15	

3.1.4 Machine Setup

The arrangement of experimental setup and analysis for milling machine, fixation of machine and workpiece, surface roughness and measurement instrument equipment are shown in Fig. 3.4a–d, respectively. Every trimming experiment was performed randomly and employed cutting parameters value via presetting condition given by the design matrix in order to avoid systematic errors. A CNC milling machine, Mori Seiki NV4000 DCG, was used for edge trimming CFRP/Al2024 composite, and its specification is presented in Table 3.2.

Figure 3.4a shows the marking on the specimens. Each specimen was labeled with specimen number, cutting speed, depth of cut, and feed rate as a reference. The specimens with the size of 100 mm × 100 mm × 4 ± 0.5 mm were made by the combination of carbon fiber-reinforced polymer (CFRP) and aluminum alloy Al2024. They were marked according to the experiment conducted and checked several times. This is to ensure the reliability of trimming process before further surface roughness checking for both CFRP and Al2024.

Figure 3.4b shows the special jig used to clamp specimen. The special jig is needed to fabricate as the specimen is thin, and it was not supposed to vibrate during trimming process. The size of the jig was 180 × 130 × 22 mm. A special jig was necessary in this process to make sure that the trimming process could be done paralleled with

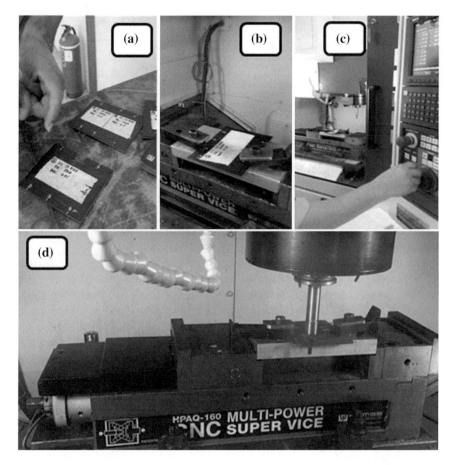

Fig. 3.4 Experimental setup, **a** marking, **b** fixing, **c** programing, and **d** edge trimming

Table 3.2 Milling machine specifications

Description	Range
Spindle speed	30,000 rpm
Table loading capacity	350 kg
Spindle drive motor	18.5/15/11 kW
Rapid traverse rate	X/Y/Z:4200 mm/min
Accuracy positioning	±0.002 mm

trimming edges of specimen. A small gap about 20 mm from the edge of jig was allowed for an overhang of the specimen and to avoid cutting jig body during each trimming pass. In addition to this, parallel bar was also used at the sides of the panel close to the edge being trimmed to securely tighten the panels without damaging the specimen.

Table 3.3 Actual design layout for experimental design

Name	Units	Low	High
Spindle speed	rpm	1000	13,500
Feed rate	mm/min	500	1000
Depth of cut	mm	0.01	0.5

As can be seen in Fig. 3.4c, the alignment and straightness of the clamping specimen were done using dial gauge indicator, followed by zero setting of x-, y-, and z-axis. The type of cutting arrangement was up milling. The experiment for a given combination of spindle speed, feed rate, and depth of cut consisted of a total of eight cutting passes along the edge of a panel of CFRP/Al2024 which was keyed in as input for machining parameters. The cutting was stopped at designated stops, and the measurements of surface roughness were made. Each experiment was repeated two times, and the results for the two experiments were averaged.

The trimming process was continued on each of the specimens as shown in Fig. 3.4d using a special trimming tool known as Kennametal (trimming interlocking bit). The edge trimming operation was carried out on CNC machining center (Mori Seiki NV4000 DCG) by varying cutting speed and feed rate by two levels and keeping depth of cut at 0.01 mm and 0.5 mm. The levels of spindle speed and feed rate were (1000, 3500 rpm) and (500, 1000 mm/min), respectively. Based on the setting of cutting parameters, the specimen was trimmed twice to ensure smooth cutting surface. The tool was changed for every four pieces of specimens due to tool wear. A worn out tool can cause a lot of burrs and affects the surface finish. In this work, edge trimming of CFRP/Al2024 specimen was conducted to determine the effect of machining quality on the surface. The machining processes were performed under dry conditions as trimming tool used in this experiment is double tooth cutter with cutting edges of PCD. The size of chip removed by the trimming tool bit was very small which is similar to grinding or cutting with abrasive cutters; thus, careful machining parameter needs to be taken.

3.1.5 Control Parameters and Experimental Design Layout

There were three machining parameters identified as process parameters for this research. They were spindle speed, feed rate, and depth of cut, each at two levels that affected the surface of the machined composite material. The responses considered here were surface roughness for CFRP and Al2024. The machining condition and their levels used for carrying out trimming operations on CFRP/Al 2024 composite are presented in Table 3.2. Based on the calculation and literature review, we come out with the average range of spindle speed, feed rate, and depth of cut (Table 3.3).

3.1.6   Edge Trimming Parameters

The project undertook trimming operation using input cutting parameters such as spindle speed, feed rate, and depth of cut, and the response parameters are surface roughness of CFRP (Ra CFRP) and aluminum (Ra Al2024). The most influenced machining parameter in composites machining is feed rate [5, 6]. By increasing the feed, more heat may be generated between tool and workpiece which can cause tool wear. When low feed rate was applied, this will give preferences to incomplete machining of fiber contained in the composite. The feed had been evaluated as reasonable by experts at Composite Testing Laboratory Asia (CTLA) Sdn. Bhd., CTRM Complex, 75350, Batu Berendam, Melaka, Malaysia. The range of feed was between 500 and 1000 mm/min. Followed by spindle speed, the increment of spindle speed leads to higher surface roughness. The selected parameters covered the minimum and maximum values of spindle speed for both CFRP composite and aluminum alloys [7–9]. The range of spindle speed was between 1600 and 3500 rpm. Based on the previous studies done by Zitoune et al. [9], Shyha et al. [8], and Nurul Khairusshimaet al. [7], depth of cut played less significant role in composites machining; hence, the range of depth of cut chosen is 0.071–0.3 mm.

3.1.7   Edge Trimming Process

Cutting tool and the workpiece were fixed and mounted on a CNC milling machine, and trimming processes were carried out according to the experiment's design layout. Programming was included in the CNC milling machine for controlling the operation of trimming. Machining was performed using a CNC milling machine Mori Seiki NV4000 DCG with a maximum spindle speed of 30,000 rpm. The parameters defined in the CNC machine were spindle speed, feed rate, and depth of cut. All trimming processes were conducted under dry cutting conditions. Two solid carbide tools of 6 mm diameter were used to carry out the full experimental design. Down milling method was chosen in order to produce a minimum surface roughness. For this experimental design, 16 specimens were prepared according to the standard 2^3 factorial design with 1 replication. The replication was run in different times spaced throughout the experiment and not as consecutive runs. The distance of machining was set to travel up to 100 mm for every setting.

3.1.8   Collecting Data of Surface Roughness

The surface roughness was measured once the machining operation was completed. Surface roughness tester equipment, Mitutoyo SJ-301, was used to measure surface roughness value for CRFP/AL 2024 with a sampling length (cut off) of 0.8 mm as shown in Table 3.4 (Fig. 3.5).

Surface roughness was the most important technical requirement in manufacturing for the best surface quality and functional behavior of the components. The distance of machining was set to travel up to 100 mm for every setting. Machined surfaces were analyzed using Mitutoyo SJ-301 surface roughness tester equipment. Five readings of machined surface were measured along 100 mm and averaged in order to encounter for an unknown error. To reduce the effect of tool wear, four burr-style router helix 15° tools had been used to trim five specimens each. Figure 3.6 shows the fixation of surface roughness tester, and Fig. 3.7 shows the setting up of workpiece for surface

Table 3.4 Specification of surface roughness tester

Technical data	Range
Measuring range	Z-axis: 350 mm
	Y-axis: 12.5 mm
Material of stylus	Diamond
Stylus tip radius	5 μm
Roughness standard	ISO
Sampling length (mm)	0.08, 0.25, 0.8, 2.5, 8
Cut-off length (mm)	0.08, 0.25, 0.8, 2.5, 8

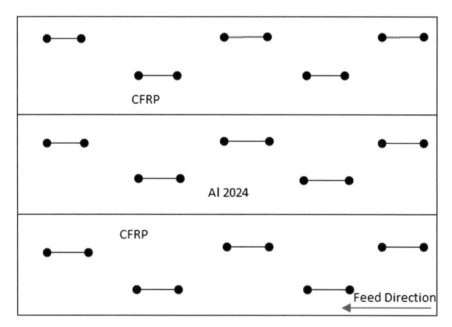

Fig. 3.5 Diagram for the measurement with Mitutoyo SJ-301

roughness data collection. Figure 3.8 illustrates the process of measuring machined surface in action.

3.1.9 DoE Approach

During the construction of the design in Design-Expert software, researchers need to define the problems and specify the information needed such as the variables, levels, and replication. After the problems have been properly defined, they should select the independent variables with their limits and the dependent variables (measured response). Independent variables of factors are set at specific values (level) and controlled in the experiment design. Then, at the end of this step the design will be computer generated, the list of experimental runs will be randomized, and experimental setting will be described. Once the experiments are systematically determined, the experiments can then be conducted based on predefined specification and criteria. At this stage, data will be measured and recorded. After all data have been recorded, the analysis of data begins. This will be done using Design-Expert software, where

Fig. 3.6 Fixation of surface roughness

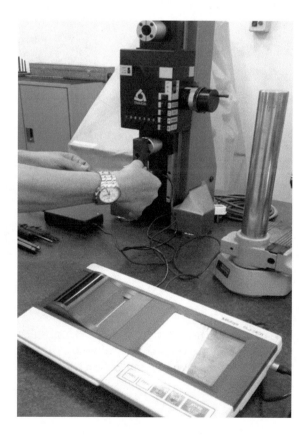

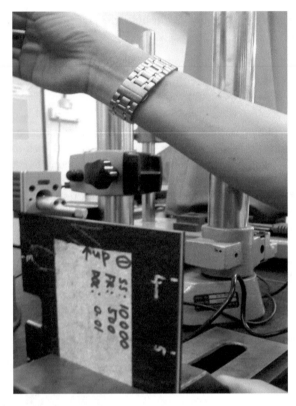

Fig. 3.7   Clamping workpiece in the fixture for surface measurement

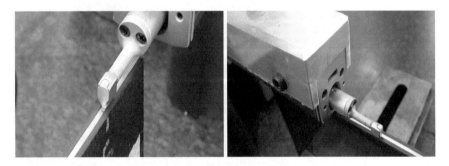

Fig. 3.8   Process of measuring of surface roughness using surface roughness tester

this software will predict which factors and interaction are significant for a particular response. Lastly, the researchers need to do a wide range of analytical and graphical techniques for model fitting and interpretation and finally document all the results and findings.

3.1.10 Design of Experiment Software

Design-Expert is a software which helps researchers to design the experiment and interpretation of the multifactor of experiments. Several design methods can be utilized using Design-Expert software such as Combined, Mixture, RSM (Box–Behnken, central composite design) and factorial designs (two-level factorial, general factorial, Taguchi). This software has the capability of fixing a more flexible model and generates D-optimal design where the other designs are not applicable. As shown in Fig. 3.9, a regular two-level factorial was involved in experimentation software.

We created a two-level 3-factor design (2^3 design); the color of cell which turned black indicates that software had chosen this design for conducting the experiment and for further analysis. From drop-down menu, replication was set by default (one replicates) and three center points were added to double check whether the developed model had a tendency to be quadratic (nonlinear) component in the relationship between a factor and a dependent variable. After that, the process continued with the next step where the software displayed screen asking the user to enter details of input variables, their unit, their types (numeric or text), and their levels as shown in Table 3.5.

The next step as shown in Fig. 3.10 required the user to deal with response variables. Here, two responses needed to be defined. One representing CFRP and

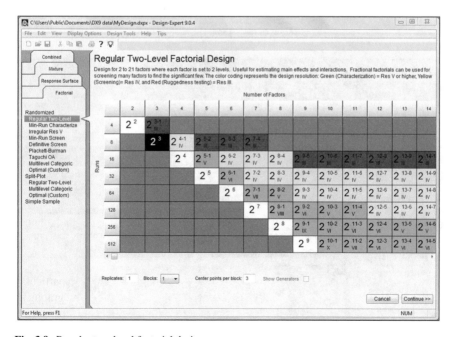

Fig. 3.9 Regular two-level factorial design

another one was for Al 2024 together with their unit of measurement. In case of more than one response needed, the user may key in number representing the total number of responses required. For example, the number 3 means three responses will be dealt with when analyzing the responses.

Table 3.6 shows that 11 experiments had been taken by the software (eight factorial experiments plus three center points) with a blank column to enter response value measured after conducting the experiments. The design had been created, and the experiment sequence had been randomized by the software.

As shown in Table 3.7, we keyed in the response value (measured surface roughness) of each experiment in the corresponding blank cell in the response column (Response 1 Ra CFRP and Response 2 Ra Al2024, respectively). Since the surface roughness for CFRP and Al2024 was recorded as two different responses, the analysis was conducted separately. The purpose of this analysis was to investigate the influence, and the interaction of machining parameters significantly affects the performance of the surface quality.

Performing analysis at this stage was conducted using ANOVA. The initial result showed that the model was insignificant for both responses (Ra CFRP and Ra Al2024) as shown in Tables 3.8 and 3.9, respectively. Most of the effects were also insignif-

Table 3.5 Three control parameters and their limits

Name	Units	Low	High
Spindle speed	rpm	1000	13,500
Feed rate	mm/min	500	1000
Depth of cut	mm	0.01	0.5

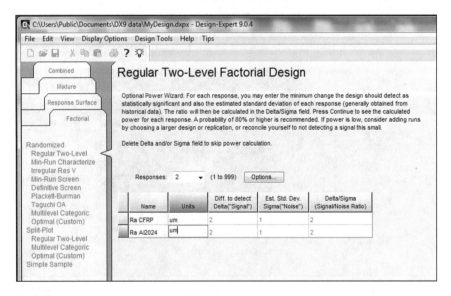

Fig. 3.10 Response parameters and their levels

Table 3.6 Design matrix

Standard no.	Run no.	Spindle speed, N (rpm)	Feed rate, fr (mm/min)	Depth of cut, dc (mm)
1	4	10,000	500	0.01
2	10	13,500	500	0.01
3	2	10,000	1000	0.01
4	6	13,500	1000	0.01
5	7	10,000	500	0.50
6	5	10,000	1000	0.01
7	8	13,500	500	0.50
8	1	13,500	500	0.01
9	9	11,750	750	0.255
10	11	11,750	750	0.255
11	3	11,750	750	0.255

Table 3.7 Experimental results of surface roughness (Ra) of CFRP and Al2024

Standard no.	Run no.	Spindle speed, N (rpm)	Feed rate, fr (mm/min)	Depth of cut, dc (mm)	Surface roughness, Ra (μm)	
					CFRP	Al2024
1	4	13,500	500	0.01	0.44	0.28
2	10	13,500	500	0.01	0.45	0.34
3	2	10,000	1000	0.01	1.04	0.21
4	6	13,500	1000	0.01	0.89	2.02
5	7	10,000	500	0.50	0.66	0.35
6	5	13,500	500	0.50	0.42	0.73
7	8	10,000	1000	0.50	0.73	0.54
8	1	13,500	1000	0.50	0.56	0.34
9	9	11,750	750	0.255	0.49	0.22
10	11	11,750	750	0.255	0.49	0.22
11	3	11,750	750	0.255	0.50	0.20

icant. However, both responses showed the source of variation for curvature was significant, which means that there was a presence of curvature. The result also showed that there was a significant lack of fit in the design. On the investigation, it was found that the addition of curvature terms in the model would improve the model sustainability . Since the curvature was also present in the interaction graph

Table 3.8 Analysis of variance results for CFRP

Source of variation	Sum of square (SS)	dof	Mean square = SS/dof	F value	P value	Remarks
Model	0.36	7	0.052	2.93	0.2031	Not significant
A-spindle speed, N	0.038	1	0.038	2.15	0.2389	
B-FEED RATE, fr	0.20	1	0.20	11.10	0.0446	
C-depth of cut, dc	0.025	1	0.025	1.44	0.3164	
AB	1.012E−003	1	1.012E−003	0.058	0.8258	
AC	9.112E−003	1	9.112E−003	0.52	0.5237	
BC	0.086	1	0.086	4.90	0.1138	
ABC	6.612E−003	1	6.612E−003	0.38	0.5831	
Residual	0.053	3	0.018			
Lack of fit	0.053	1	0.053	1581.01	0.0006	Significant
Pure error	6.667E−005	2	3.333E−005			

Table 3.9 Analysis of variance results for Al2024

Source of variation	Sum of square (SS)	dof	Mean square = SS/dof	F value	P value	Remarks
Model	2.49	7	0.36	23.25	0.1808	Not significant
A-spindle speed, N	0.53	1	0.53	4.80	0.1163	
B-feed rate, fr	0.25	1	0.25	2.27	0.2291	
C-depth of cut, dc	0.099	1	0.099	0.90	0.4119	
AB	0.17	1	0.17	1.56	0.3000	
AC	0.36	1	0.36	3.26	0.2688	
BC	0.41	1	0.41	3.74	0.1486	
ABC	0.68	1	0.68	6.20	0.0886	
Residual	0.033	3	0.11			
Lack of fit	0.033	1	0.053	2462.39	0.0004	Significant
Pure error	2.667E−004	2	1.333E−004			

(Fig. 3.11), it became clear that the full factorial design may not be capable to estimate the curvature. Therefore, there is a need to augment the existing design to central composite design (CCD) to estimate the quadratic term.

Generally, there were two approaches to overcome this situation: firstly, by restarting completely with a new experiment applying CCD approach; secondly, by using

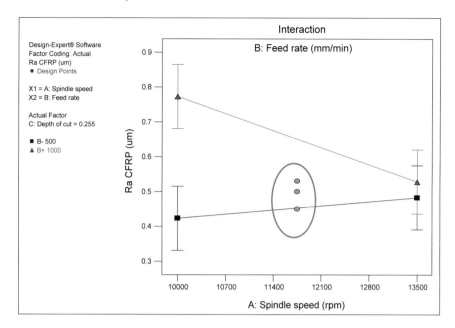

Fig. 3.11 Interaction graph of CFRP

the existing design data with additional run called center composite design (CCD) as suggested by Lalwani et al. [10] and Mahdi et al. [11]. This second approach is an extension of two-level full factorial designs. Since the cost of test panel was relatively expensive and consumed a lot of time for data collection, we decided to go with the second option.

3.1.11 Augment Design

Design-Expert software provides the facilities to modify the existing design to central composite design. The purpose of Augment Design was to resolve the ambiguities that resulted from a previous design by conducting further experimental runs. With existing design table, the Augment Design platform constructs additional runs in a way that optimizes the overall design process.

As shown in Fig. 3.12, users need to go to menu "Design Tools">Augment Design>Augment. Then, they need to select central composite design from drop-down menu (Fig. 3.13.) and click OK. They need to enter 1 run per axial points and 2 in additional center points (Fig. 3.14) and choose face centered as illustrated in Fig. 3.15, and then click OK. Design-Expert provided the Augment Design with eight trials points added in the original design for each response to make it central composite design as shown in Table 3.10.

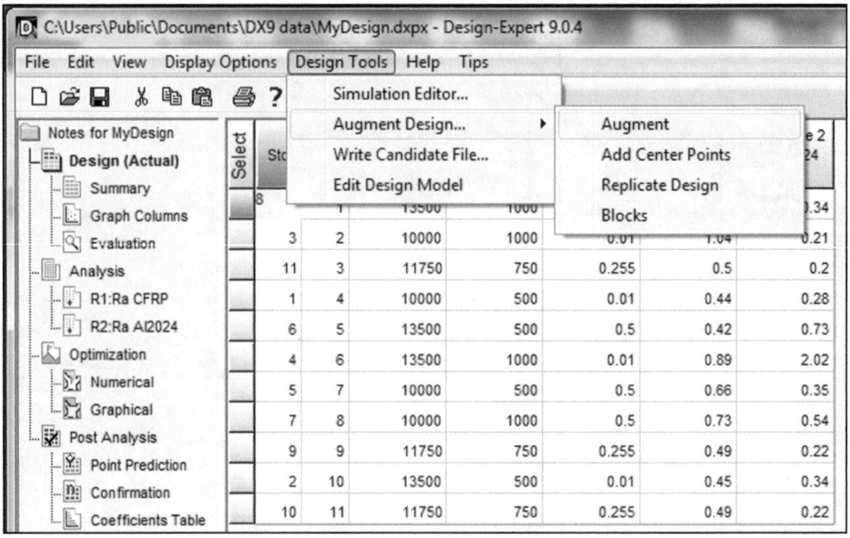

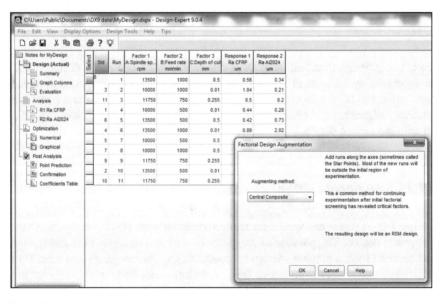

Fig. 3.12 Augment design tools interface

Fig. 3.13 Factorial design augmentation

3.1.12   Validation of Experiments

If the predicted distribution of target responses in a developed mathematical model matched observational data, the model may be considered as verified. In order to

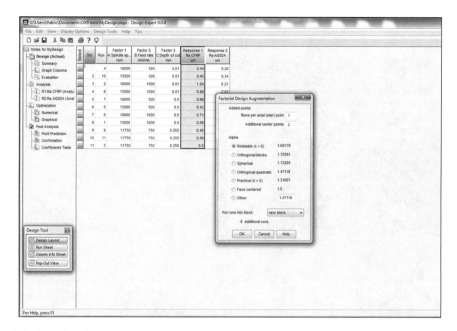

Fig. 3.14 Entering axial point and center point

verify the models, normally three confirmation experiments are used to perform the test [4]. The tests were performed using two proposed solutions which were selected at random from the Numerical Optimization Report. The trimming process is conducted using the same procedure settings as described earlier in this chapter. Three measured results for each response were compared with the predicted value using the mathematical model developed. It was helpful to know what percent of predicted values differ from measured values. The comparison was based on the following formula:

$$\text{Percentage Error} = \frac{\text{Predicted Value} - \text{Measured Value}}{\text{Measured Value}} \times 100\% \qquad (3.1)$$

In most of the engineering cases, if the percentage error or difference between predicted and measured value was less than 10%, the developed model will be accepted. This calculation helps to evaluate the relevance of the experimental results. If the comparison showed a difference of more than 10%, it is highly likely that some mistakes have occurred and the experiment should be reviewed to find the source of the error to be corrected.

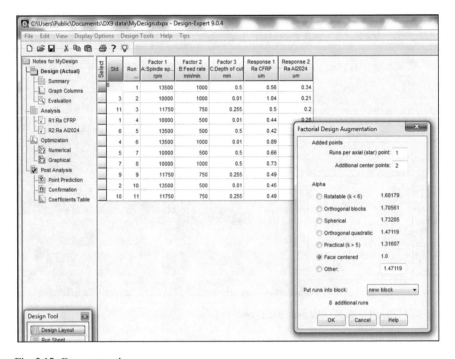

Fig. 3.15   Face centered

3.1.13   Summary of DoE Approach

Design of experiments is a series of tests in which purposeful changes are made
to the input variables of a system or process, and the effects on response variables
are measured. Design of experiments is applicable to both physical processes and
computer simulation models. Experimental design is an effective tool for maximizing
the amount of information gained from a study while minimizing the amount of
data to be collected. Factorial experimental designs investigate the effects of many
different factors by varying them simultaneously instead of changing only one factor
at a time. Factorial designs allow estimation of the sensitivity to each factor and also
to the combined effect of two or more factors.

The nine main steps to implement an experimental design are as follows:

(i)   State the objective of the study.
(ii)   Determine the response variable(s) of interest that can be measured.
(iii) Determine the controllable factors of interest that might affect the response
       variables and the levels of each factor to be used in the experiment.
(iv)   Determine the total number of runs in the experiment, ideally using estimates
       of variability, precision required, size of effects expected, etc., but more likely

Table 3.10 Augment design (extension of run)

Standard no.	Block no.	Run no.	Spindle speed, N (rpm)	Feed rate, fr (mm/min)	Depth of cut, dc (mm)	Surface roughness, Ra (μm)	
						CFRP	Al2024
1	1	4	13,500	500	0.01	0.44	0.28
2	1	10	13,500	500	0.01	0.45	0.34
3	1	2	10,000	1000	0.01	1.04	0.21
4	1	6	13,500	1000	0.01	0.89	2.02
5	1	7	10,000	500	0.50	0.66	0.35
6	1	5	13,500	500	0.50	0.42	0.73
7	1	8	10,000	1000	0.50	0.73	0.54
8	1	1	13,500	1000	0.50	0.56	0.34
9	1	9	11,750	750	0.255	0.49	0.22
10	1	11	11,750	750	0.255	0.49	0.22
11	1	3	11,750	750	0.255	0.50	0.20
12	2	16	10,000	750	0.25		
13	2	17	13,500	750	0.25		
14	2	15	11,750	500	0.25		
15	2	18	11,750	1000	0.25		
16	2	12	11,750	750	0.01		
17	2	19	11,750	750	0.50		
18	2	13	11,750	750	0.25		
19	2	14	11,750	750	0.25		

based on the available time and resources. Design the experiment, remembering to randomize the runs.

(v) Perform the experiment strictly according to the experimental design, including the initial setup for each run in a physical experiment. Do not swap the run order to make the job easier.

(vi) Analyze the data from the experiment using the analysis of variance method.

(vii) Interpret the results, and state the conclusions in terms of the subject matter.

(viii) Consider performing a second confirmatory experiment if the conclusions are very important or are likely to be controversial.

(ix) Document and summarize the results and conclusions, in tabular and graphical form, for the report or presentation on the study.

References

1. Dandekar, C. R., & Shin, Y. C. (2012). Modeling of machining of composite materials: A review. *International Journal of Machine Tools and Manufacture, 57,* 102–121. https://doi.o rg/10.1016/j.ijmachtools.2012.01.006.
2. Teti, R. (2002). Machining of composite materials. *CIRP Annals-Manufacturing Technology, 51*(2), 611–634. https://doi.org/10.1016/s0007-8506(07)61703-x.
3. Davim, J. P., & Reis, P. (2005). Damage and dimensional precision on milling carbon fiber-reinforced plastics using design experiments. *Journal of Materials Processing Technology, 160,* 160–167. https://doi.org/10.1016/j.jmatprotec.2004.06.003.
4. Noordin, M. Y., Venkatesh, V. C., Sharif, S., Elting, S., & Abdullah, A. (2004). Application of response surface methodology in describing the performance of coated carbide tools when turning AISI 1045 steel. *Journal of Materials Processing Technology, 145,* 46–58. https://do i.org/10.1016/S0924-0136(03)00861-6.
5. Iliescu, D., Gehin, D., Gutierrez, M. E., & Girot, F. (2010). Modeling and tool wear in drilling of CFRP. *International Journal of Machine Tools and Manufacture, 50*(2), 204–213. https:// doi.org/10.1016/j.ijmachtools.2009.10.004.
6. Zitoune, R., Krishnaraj, V., Sofiane, B., Collombet, F., Sima, M., & Jolin, A. (2012). Composites: Part B influence of machining parameters and new nano-coated tool on drilling performance of CFRP/Aluminium sandwich. *Composites: Part B, 43*(3), 1480–1488. https:// doi.org/10.1016/j.compositesb.2011.08.054.
7. Nurul Khairusshima, M. K., Che Hassan, C., Jaharah, A., & Nurul Amin, K. (2012). Tool wear and surface roughness on milling carbon fibre reinforced plastic using chilled air. *Journal of Asian Scientific Research, 2*(11), 593–598.
8. Shyha, I., Soo, S. L., Aspinwall, D. K., Bradley, S., Dawson, S., & Pretorius, C. J. (2010). Drilling of Titanium/CFRP/Aluminium stacks. *Key Engineering Materials, 448,* 624–633. https://doi.org/10.4028/www.scientific.net/KEM.447-448.624.
9. Zitoune, R., Krishnaraj, V., & Collombet, F. (2010). Study of drilling of composite material and aluminium stack. *Composite Structures, 92*(5), 1246–1255. https://doi.org/10.1016/j.com pstruct.2009.10.010.
10. Lalwani, D. I., Mehta, N. K., & Jain, P. K. (2007). Experimental investigations of cutting parameters influence on cutting forces and surface roughness in finish hard turning of MDN250 steel. *Journal of Materials Processing Technology, 6,* 167–179. https://doi.org/10.1016/j.jma tprotec.2007.12.018.
11. Mahdi, S., Shahabadi, S., & Reyhani, A. (2014). Optimization of operating conditions in ultra-filtration process for produced water treatment via the full factorial design methodology. *Separation and Purification Technology, 132,* 50–61. https://doi.org/10.1016/j.seppur.2014.04.051.

Chapter 4
Learning from the Carbon Fiber-Reinforced Plastic Project

4.1 Surface Roughness (Ra) of CFRP and Al2024

Design-Expert V9 software was used to predict the cutting parameters that had an important effect for a particular response. Since the surface roughness for CFRP and Al2024 was recorded as two different responses, the analysis was conducted separately. The obtained results from the experiment are listed in Table 4.1 which is then used for further analysis using Design-Expert V9 software. The input variables were spindle speed, feed rate, and depth of cut; meanwhile, the responses were surface roughness for CFRP and Al2024. The analysis began with the analysis of CFRP machined surface and then followed by Al2024 machined surface.

Table 4.1 shows the results of the surface roughness when machining CFRP/Al2024 under dry machining condition. Through a range of preset parameters, the results of surface roughness were recorded from 0.42 to 1.04 μm for CFRP and 0.2 to 0.8 μm for Al2024.

Surface roughness presented in Table 4.1 is based on the average of five readings in order to encounter for an unknown error. For example, experiment no. 17, surface roughness of CFRP, i.e., 0.53 μm was the average of five (5) readings, 0.51, 0.51, 0.58, 0.48, and 0.59 μm.

4.2 Analysis of Variance of CFRP and Al2024 Response

ANOVA was performed to analyze the factors that affect the surface roughness as shown in Tables 4.2 and 4.3. Most of the terms have the same value P (<0.05); a significant order of each term was prepared in accordance with the greatest F value. The analysis of variance (ANOVA) for surface roughness of CFRP and Al2014 was conducted separately. Statistically, the quadratic model F value of 6.29 implied that

© The Author(s), under exclusive license to Springer Nature Singapore Pte Ltd 2019
S. B. Mohamed et al., *Down Milling Trimming Process Optimization for Carbon Fiber-Reinforced Plastic*, SpringerBriefs in Applied Sciences and Technology,
https://doi.org/10.1007/978-981-13-1804-7_4

Table 4.1 Experimental results of surface roughness (Ra) of CFRP and Al2024

Experiment no.	Spindle speed, N (rpm)	Feed rate, fr (mm/min)	Depth of cut, dc (mm)	Surface roughness, Ra (μm)	
				CFRP	Al2024
1	13,500	1000	0.50	0.56	0.34
2	10,000	500	0.05	0.66	0.35
3	10,000	1000	0.50	0.73	0.54
4	11,750	750	0.25	0.49	0.22
5	13,500	500	0.01	0.45	0.34
6	10,000	1000	0.01	1.04	0.21
7	13,500	500	0.50	0.42	0.73
8	10,000	500	0.01	0.44	0.28
9	11,750	750	0.25	0.49	0.22
10	13,500	1000	0.01	0.89	2.02
11	11,750	750	0.25	0.50	0.20
12	11,750	1000	0.25	1.03	0.55
13	11,750	750	0.25	0.75	0.37
14	11,750	750	0.50	0.54	0.26
15	11,750	500	0.25	0.50	0.42
16	11,750	750	0.01	0.43	0.47
17	13,500	750	0.25	0.53	0.80
18	11,750	750	0.25	0.74	0.36
19	10,000	750	0.25	0.82	0.74

the model was significant. P value is less than 0.05, which indicated that the model was significant. In this case A, B and BC were significant model terms. The analysis showed that spindle speed and feed rate had a significant effect on surface roughness of CFRP. The depth of cut had no significant effect, but there was a significant interaction effect between feed rate and depth of cut as shown in Table 4.2.

The main effect and interaction of cutting parameter were further analyzed by means of response graph. Figure 4.1 shows the influence of feed rate on machined surface of CFRP at two different settings of B-feed rate. At low B-feed rate (500 mm/min) and low C-depth of cut (0.5 mm), the influence on machine surface was at minimum when compare to high C-depth of cut. Critically, increasing the depth of cut produces high cutting forces and vibrations on the workpiece as well as cutting tools which eventually had a large effect on surface finish.

Analysis of variance for AL2024 is shown in Table 4.3. It can be clearly seen that the quality of surface roughness of Al2024 was influenced by the interaction between A-spindle speed and B-feed rate. The results of the model P value of 0.0384 and lack of fit of 0.0088 were significant. The C-depth of cut of 0.0114 had a significant effect on surface roughness of Al2024, but A-spindle speed of 0.1068 and the B-feed rate of 0.2531 were not significant. The interaction effect is shown using the interaction graph in Fig. 4.2.

Table 4.2 Analysis of variance for CFRP (Augment)

Source of variation	Sum of square (SS)	d.o.f	Mean square = SS/d.o.f	F value	P value	Remarks
Model	0.63	9	0.070	6.29	.0182	Significant
A-spindle speed, N	0.091	1	0.091	8.19	0.0287	Significant
B-feed rate, fr	0.25	1	0.25	22.72	0.0031	Significant
C-depth of cut, dc	0.055	1	0.055	4.99	0.0669	
AB	1.513E−003	1	1.513E−003	0.14	0.7243	
AC	0.017	1	0.017	1.49	0.7243	
BC	0.11	1	0.11	9.74	0.0206	Significant
A^2	0.010	1	0.010	0.95	0.3685	
B^2	0.019	1	0.019	1.73	0.2366	
C^2	0.059	1	0.059	5.36	0.0598	
Residual	0.066	6	0.011			
Lack of fit	0.066	4	0.017	497.29	0.0020	Significant

Std. dev. = 0.11; mean = 0.66; R-squared = 0.9042; pred. R-squared = −0.0474; adj. R-squared = 0.7605

Table 4.3 Analysis of variance for Al2024 (Augment)

Source of variation	Sum of square (SS)	d.o.f	Mean square = SS/d.o.f	F value	P value	Remarks
Model	0.42	9	0.046	4.60	0.0384	Significant
A-spindle speed, N	0.036	1	0.036	3.59	0.1068	
B-feed rate, fr	0.16	1	0.16	1.60	0.2531	
C-depth of cut, dc	0.13	1	0.13	12.94	0.0114	Significant
AB	0.061	1	0.061	6.11	0.0484	Significant
AC	3.966E−004	1	3.966E−004	0.040	0.8489	
BC	4.561E−004	1	4.561E−004	0.045	0.8382	
A^2	0.060	1	0.060	6.00	0.0498	
B^2	0.021	1	0.021	2.11	0.1962	
C^2	0.017	1	0.017	1.74	0.2354	
Residual	0.060	6	0.010			
Lack of fit	0.060	4	0.015	112.32	0.0088	Significant

Std. dev. = 0.11; mean = 0.40; R-squared = 0.8735; pred. R-squared = −1.6134; ajd. R-squared = 0.6837

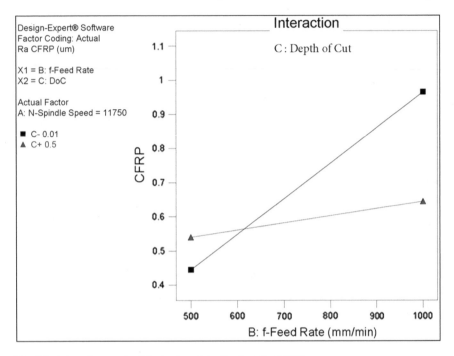

Fig. 4.1 Interaction graph of feed rate and depth of cut for CFRP

The response graph for Al2024 shows the influence of feed rate on machined surface of CFRP at two different settings of *B*-feed rate. At high *B*-feed rate (1000 mm/min) and high *A*-spindle speed (13,500 rpm), the influence on machined surface was at maximum when compared to low *B*-feed rate. However, at the low *B*-feed rate and high *A*-spindle speed, the effect was not so significant. This is due to the residual stress distribution effects of cutting speed for Al2024. These residual stress values increased or decreased and fluctuated along the measured surfaces and could be related to the coarse grain microstructure of the Al2024.

4.3   Final Equation in Terms of Actual Factors for CFRP and Al2024 Responses

The estimated value of the surface roughness took into account all the factors and interactions among other factors, the custom for surface roughness obtained through quadratic equations. A mathematical equation quadratic model was developed to estimate the surface roughness by all significant factors. This equation was generated using the software Design-Expert.

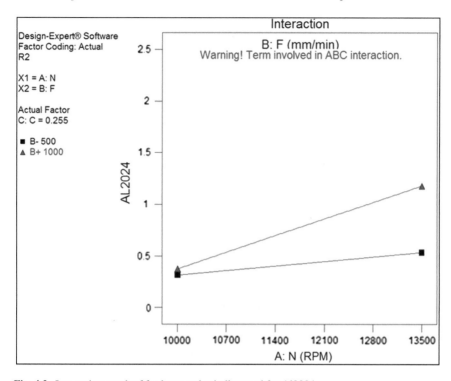

Fig. 4.2 Interaction graph of feed rate and spindle speed for Al2024

$$
\begin{aligned}
\text{Ra CFRP} = &+ 2.90628 - 5.72890e{-}004 * \text{Spindle speed} + 2.95541e{-}003 \\
&* \text{Feed rate} + 0.64861 * \text{Depth of cut} + 3.14286e{-}008 \\
&* \text{Spindle speed} * \text{Feed rate} - 1.06018e{-}004 \\
&* \text{Spindle speed} * \text{Depth of cut} - 1.89455e{-}008 \\
&* \text{Feed rate} * \text{Depth of cut} + 2.22111e{-}008 * \text{Spindle speed}^2 \\
&- 1.47165e{-}006 * \text{Feed rate}^2 + 3.29578 * \text{Depth of cut}^2 \quad (4.1)
\end{aligned}
$$

$$
\begin{aligned}
\text{Ra Al2024} = &+ 4.78416 - 1.07193e{-}003 * \text{Spindle speed} \\
&+ 4.54534e{-}0038 * \text{Feed rate} - 0.49480 * \text{Depth of cut} \\
&- 2.00000e{-}007 * \text{Spindle speed} * \text{Feed rate} + 1.64204e{-}005 \\
&* \text{Spindle speed} * \text{Depth of cut} - 1.23272e{-}004 * \text{Feed rate} \\
&* \text{Depth of cut} + 5.32785e{-}008 \\
&* \text{Spindle speed}^2 - 1.54935e{-}006 * \text{Feed rate}^2 \quad (4.2)
\end{aligned}
$$

To validate the effectiveness of the model, the value of factors was included in Eqs. (4.1) and (4.2). The difference between the values measured during the actual and the predicted values of the model is set out in Table 4.4 with an average error

Table 4.4 Validation test on surface roughness

No.	Spindle speed, N (rpm)	Feed rate, f_r (mm/min)	Depth of cut, d_c (mm)	Surface roughness, Ra (μm)						
				CFRP			Al2024			
				Actual	Predict	% error	Actual	Predict	% error	
1	11,750	750	0.255	0.594	0.607	2.20	0.320	0.335	4.81	
2	11,750	754	0.257	0.586	0.610	4.03	0.328	0.335	2.04	
Average % error						3.11			3.43	

of 3.11% for CFRP and 3.43% for Al2024. This model is good and acceptable because the calculated error is ±10% [1]. The validation experiment was conducted at optimum level of recommended control factors setting to compare the deviation of predicted value from actual/measured value.

4.4 The Form of Interaction in a 3D Graph of CFRP Response

From the developed mathematical model, the effect of trimming parameters on surface roughness was observed. Figure 4.3 shows the interaction feed rate and depth of cut to the surface roughness of CFRP. It was found that an extreme form of the curve which showed the interaction of these two factors is significant. This is consistent with the fact that the interaction of depth of cut and feed rate is a significant factor for determining the quality of the machined surface as presented in ANOVA Table 4.2.

From the graph, the higher the feed rate, the higher the value of surface roughness. The highest value is at $B = 1000$ mm/min and $C = 0.01$ mm. X1 represents the feed rate and depth of cut is represented by X2 at a constant spindle speed of $A = 11{,}702.7$ rpm. Range of the region for surface roughness of CFRP is between 0.42 and 1.04 μm. The curve shape shows that the interaction between feed rate and depth of cut is significant. However, when the feed rate is reduced, it increases the roughness of machined surface.

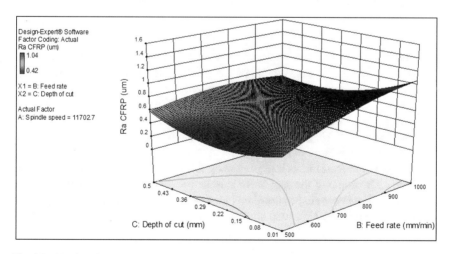

Fig. 4.3 3D plot of interaction between feed rate and depth of cut

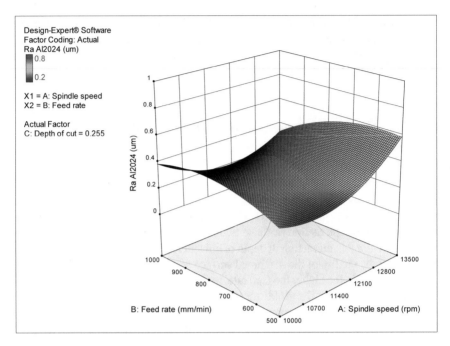

Fig. 4.4 3D plot of interaction between spindle speed and feed rate

4.5 The Form of Interaction in a 3D Graph of Al2024 Response

The interaction effect between spindle speed and feed rate is shown in Fig. 4.4. An extreme form of the curve was found in this graph, which showed the interaction of these two factors is significant. This is consistent with the fact that the interaction of depth of cut and feed rate is a significant factor for determining the quality of the machined surface as presented in Table 4.3.

From the graph, the higher the feed rate, the higher the value of surface roughness. The highest value is at $A = 13,500$ rpm and $B = 500$ mm/min. X1 represents the spindle speed and feed rate is represented by X2 at a constant depth of cut of $C = 0.255$ mm. The range of the region for surface roughness of CFRP is between 0.2 and 0.8 μm. The curve shape shows that the interaction between spindle speed and feed rate is significant. However, when the spindle speed is increased at high feed rate, it increases the roughness of machined surface.

Fig. 4.5 Optimization steps

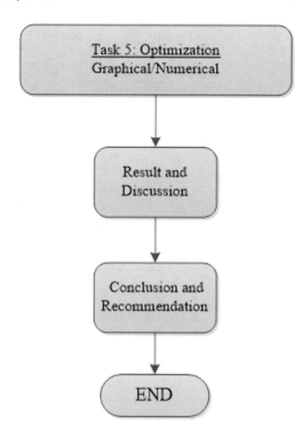

4.6 Edge Trimming Parameter Optimization

In Design-Expert software V9, the optimization part searches for a combination factor level that simultaneously satisfies the requirements placed. Numerical and graphical optimization methods were used in this work by selecting the desired goals for each factor and response. The optimization process starts by using the mathematical model developed by Design-Expert software. This section starts by describing the procedure undertaken in performing the numerical optimization and followed by graphical optimization as illustrated in Fig. 4.5.

4.6.1 Numerical Optimization

Design-Expert allows criteria setting for all variables including factors and desired responses. In this case, the researcher left the setting of spindle speed, feed rate, and depth of cut without any changes. The target value of response criteria was

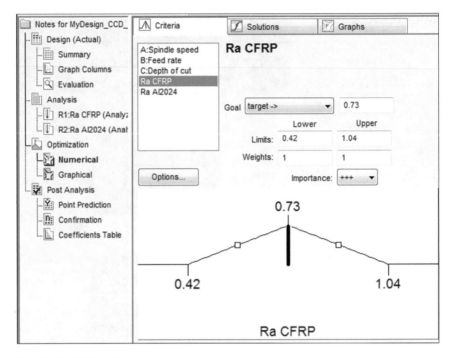

Fig. 4.6 Ra CFRP setting criteria

set to below 1 μm for Al2024 and CFRP. Since the responses of this experiment involved two responses, thus the setting of numerical criteria optimization was done separately.

For CFRP (Fig. 4.6), the settings for lower and upper bounds were 0.42 and 1.04, respectively. These limits indicated that it was the most desirable range to achieve the targeted value of less than 1 μm. These settings create the following desirability functions of Ra CFRP. From 0.42 to 0.73, desirability (di) ramps up from zero to 0.73. From 0.73 to 1.04, di ramps back down to zero. If value of responses is greater than 1.04 and less than 0.42, then the desirability (di) equals to zero meaning any values outside the stated range are not acceptable.

Meanwhile, for Al2024 (Fig. 4.7), the settings for lower and upper bounds were 0.2 and 0.8, respectively. These limits indicated that it was the most desirable range to achieve the targeted value of less than 1 μm. These settings create the following desirability functions of Ra Al2024. From 0.2 to 0.5, desirability (di) ramps up from zero to one. From 1 to 0.8, di ramps back down to zero. If value of responses is greater than 0.5 and less than 0.2, then the desirability (di) equals to zero which means any values outside the stated range are not acceptable.

The Design-Expert provides optimal designs with the different desirability factors ranging from the least desirable to the most desirable (0–1). Desirability is an objective function using a mathematical method to determine the optimum factors

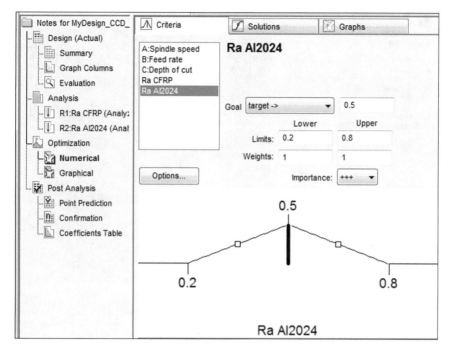

Fig. 4.7 Ra Al2024 setting criteria

Table 4.5 Numerical optimization report on five solutions

No.	Spindle speed	Feed rate	Depth of cut	Ra CFRP	Ra Al2024	Desirability
1	11,750.001	750.000	0.255	0.607	0.335	0.771
2	11,775.572	750.011	0.255	0.606	0.336	0.769
3	11,749.985	758.312	0.255	0.612	0.334	0.767
4	11,717.068	750.005	0.254	0.609	0.334	0.767
5	11,749.981	762.408	0.255	0.615	0.333	0.767

for a particular defined response. The goal of optimization is to obtain a secure set of conditions that will satisfy all the goals. The program randomly picks a set of conditions from which to begin its search for desirable outcomes. After grinding through cycles of optimization, Design-Expert sorted the results in tabular form as shown in Table 4.5. The number of possible solutions with desirability of 1 was 5 numbers of settings.

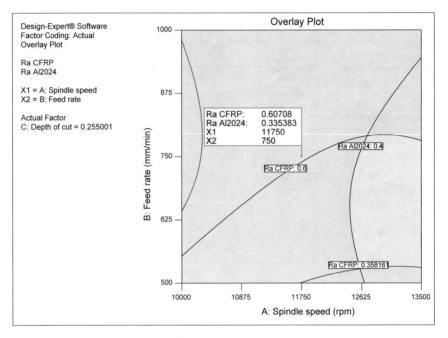

Fig. 4.8 Overlay plot

4.6.2 *Graphical Optimization*

Graphical optimization in Design-Expert defines the region where the requirements simultaneously meet the proposed criteria. Overlay region can be presented on a contour plot. This is commonly referred to as an overlay plot graph.

Graphical optimization displays the area of feasible response values in the factor space. Regions that do not fit the optimization criteria are shaded gray. Any "window" that is not gray shaded satisfies the goals for every response. By default, the area that satisfied the constraints will be yellow, while the area that does not meet the criteria is gray.

Overlay plot in Fig. 4.8 shows the graphical overview of the proposed factor settings with yellow shaded area meeting the target responses. The yellow zone represents the zone of optimum responses of surface roughness for both CFRP and Al2024 in between 0.4 and 0.6 μm (where in the objective of this research is below than 1 μm) with the setting 0.255 mm depth of cut. The range of spindle speed for a presetting surface roughness value should be between 12,300 and 13,500 rpm, and the feed rate should be approximately between 560 and 780 mm/min.

As can be predicted by the overlay plot, an estimation of surface roughness for CFRP would be 0.60708 μm and for an estimation of surface roughness for Al2024 would be 0.3354 μm. These can be obtained with the setting of the cutting parameter: X1 = 11,750 rpm, Spindle speed: X2 = 750 mm/min and the depth of cut 0.255 mm.

4.7 Conclusion and Recommendations for Future Works

Based on this research project, it can be concluded that trimming process using solid carbide end mills tools (Kennametal Burr-Style Routers Helix 15^0, KCN05 6 mm diameter) via down milling approach is efficient. Feed rate is found to be the most significant factor to the surface roughness of CFRP. The interaction among feed rate and depth of cut as well as spindle speed is deemed to have the least effect. This is because the exposure of the machined surface produced is controlled by the feed rate. The depth of cut is the most prominent factor to the surface roughness of Al2024 and the interaction between spindle speed and feed rate.

In validating the optimum cutting parameter, five optimal trimming parameters have been proposed numerically showing the response of Ra CFRP and Ra Al2024 within its limit of below than 1 μm. In optimization methodology, numerical optimization and graphical optimization can be used to identify the optimal trimming parameters in such a way to produce a good-quality machined surface in trimming of CFRP/Al2024 composite material.

For future works, the number of other response parameters such as type of cutting tool, tool wear, cutting forces, and heat generation should be considered so as to improve the machined surface quality, machining time, machine utilization heat generated. The effect of process parameters may be assessed by varying the orientation of fiber present in the composite material.

Reference

Hills, R. G., Cruces, L., & Trucano, T. G. (1999). Statistical validation of engineering and scientific models : Background. Retrieved from http://www4.ncsu.edu/~rsmith/Hills-1.pdf.

Appendix A

NV 4000 DCG

Presenting the ideal vertical machining center

High-speed and high-quality—in order to combine these conflicting factors, DMG MORI took a fresh look at the structure of machine tools. The NV4000 DCG, a high-precision vertical machining center, achieves both high speed and high quality thanks to the innovative technology.

Machine specifications

Item			NV4000 DCG	NV4000 DCG HSC	
			12,000 min⁻¹	20,000 min⁻¹	[30,000 min⁻¹]
Travel	X-axis travel <longitudinal movement of table>	mm (in.)	600 (23.6)		
	Y-axis travel <cross movement of saddle>	mm (in.)	400 (15.7)		
	Z-axis travel <vertical movement of spindle head>	mm (in.)	400 (15.7)		
	Distance from table surface to spindle gauge plane	mm (in.)	100—500 (3.9—19.7) [150—550 (5.9—21.7) <APC <raised column 100 (3.9)> specifications>]		
Table	Distance from floor surface to table surface	mm (in.)	900 (35.4) [950 (37.4) <APC specifications>]		
	Table working surface	mm (in.)	700×450 (27.6×17.7) <for APC specifications, please check the pallet configuration diagrams.>		
	Table loading capacity	kg (lb.)	350 (770) [250 (550) <APC specifications>]		
	Table surface configuration <T slots width×pitch×No. of T slots>		18 mm×100 mm×4 (0.7 in.×3.9 in.×4)		
Spindle	Max. spindle speed	min⁻¹	12,000	20,000	30,000
	Number of spindle speed ranges		1		
	Type of spindle taper hole		No. 40		No. 40 (HSK-F63)
	Spindle bearing inner diameter	mm (in.)	70 (2.8)		60 (2.4)
Feedrate	Rapid traverse rate	mm/min (ipm)	X, Y, Z: 42,000 (1,653.5)		
	Cutting feedrate	mm/min (ipm)	X, Y, Z: 1—42,000 (0.04—1,653.5) <for look-ahead control <theoretical value>>		
	Jog feedrate	mm/min (ipm)	0—5,000 (0—197.0) <20 steps>		
ATC	Type of tool shank		BT40* [DIN40] [CAT40] [HSK-A63]		HSK-F63
	Type of retention knob		DMG MORI SEIKI 90° type [45°(MAS-I)] [60°(MAS-II)] [HSK-A63]		HSK-F63
	Tool storage capacity		20 [40] [80]		
	Max. tool diameter — With adjacent tools	mm (in.)	80 (3.1) [70 (2.7) <with the 40- and 60-tool specified tool magazine>]		
	Max. tool diameter — Without adjacent tools	mm (in.)	125 (4.9)		
	Max. tool length	mm (in.)	250 (9.8)		
	Max. tool mass	kg (lb.)	8 (17.6)		3 (6.6)
	Max. tool mass moment <from spindle gauge line>	N·m (ft·lbf)	11 (8.1) <a tool with a mass moment greater than the maximum tool mass moment may cause problems during ATC operations even if it satisfies the other conditions.>		
	Method of tool selection		Fixed address, shorter route access		
	Tool changing time — Tool-to-tool	s	1.0		
	Tool changing time — MAS	s	2.8		
	• The time differences are caused by the different conditions <travel distances, etc> for each standard. • Depending on the arrangement of tools in the magazine, the cut-to-cut (chip-to-chip) time may be longer. — Cut-to-cut (chip-to-chip) <without ATC shutter> ISO 10791-9 JIS B6336-9	s	20-tool specifications: 5.5 (max.), 3.6 (min.) [40-tool specifications: 10.9 (max.), 3.6 (min.)]		
Motor	Spindle drive motor	kW (HP)	18.5/15/11 (24.7/20/15) <10 min/30 min/cont> (high-speed winding side)		18.5/13 (24.7/17.3) <1 min/cont>
	Feed motor	kW (HP)	X: 1.6 (2.1), Y: 1.6 (2.1)×2, Z: 3.0 (4.0)×2		
	Coolant pump motor (50 Hz/60 Hz)	kW (HP)	0.6 (0.8)/1.02 (1.37)		
Power sources <standard>	Electrical power supply <cont>	等価容量 (kVA)	27.7		30.0
	Compressed air supply	MPa (psi), L/min (gpm)	0.5 (72.5), 200 (52.8) <when the tool tip air blow is regularly used, air supply of more than 300 L/min (79.2 gpm) is separately required> <ANR>		

Appendix B

Mobility

- The integral battery in the SJ-301 enables surface roughness measurements to be made even in places where there is no mams connection.
- Portable and practical: The drive unit and probes are stored for transport inside the device (a carrying case is provided with the SJ-301).
- For measuring, the evaluation unit does not have to be taken out of the carrying case. An additional protective film is supplied for the display.

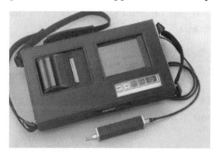

Highly sensitive measurements

- The SJ-301 detector uses the differential inductance method as used in many high-end instruments.
- Profile detail can be seen at a resolution of up to 0.01 μm (0.4 in.) in the Z-axis direction.

- Parameters requiring high-accuracy feed, such as Sm and S, can also be measured with the SJ-301.
- The probe can be retracted into the drive unit after making a measurement for maximum protection against accidental damage.

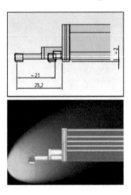

Problem-free probe installation

- Optional probes for measuring features such as small diameters or deep grooves are available as an option.
- No tool is needed when changing the probe. Just remove the probe and replace it with an optional probe.
- Using just an SJ-301, the optional accessories mean that a very wide variety of workpieces can be measured.

Adjustable evaluation length range

- The instrument can be set to measure over a length from 0.3 to 12.5 mm (0.012″ to 0.49″) in 0.1 mm increments.

Specifications SJ-301

Code No.* 178-952-2

Drive Unit	Z-axis stroke: 350 µm, X-axis feed: 12,5 mm
Measuring range	Measuring: 0,,25 mm/s, 0,5 mm/s
Speed	Returning: 1 mm/s
Cable length	1 m
Mass	190 g
Standard Probe	Code No. 178-395
Type	Differential inductance
Measuring range	350 µm (-200 µm to +150 µm)
Stylus	Diamond cone
Tip radius	2 µm
Skid radius	40 mm
Contact force	0,75 mN
Recorded profiles	Primary profile (R), Roughness profile (R), DIN EN ISO 13565-1, MOTIF.R, MOTIF.W
Roughness parrameters	Ra, Ry, Rz, Rt, Rp, Rq, Rv, Sm, S, Pc, R3z, mr, Rpk, Rvk, sc, Rk, Mr1, Mr2, Lo, Ppi, R,AR, Rx, A1, A2, Vo, HSC. mrd, sk, Ku, Da, Dq, Wte, W, AW
Analysis graphs	BAC1, BAC2, ADC
Roughness standards	JIS, DIN, ISO, ANSI
Evaluation length (L)	0,08 mm, 0,25 mm, 0,8 mm, 2,5 mm, 8 mm
Cut-off wavelengths	lc: 0,08 mm, 0,25 mm, 0,8 mm, 2,5 mm, 8 mm
	ls: 2,5 µm, 8 µm, 25 µm
No. of sampling lengths	x1, x3, x5, xL**
Digital filters	2CR, PC75 (phasen-korrigiert), Gauß
Resolution/range	0,4 µm / 350 µm, 0,1 µm / 100 µm, 0,05 µm / 50 µm, 0,01 µm/10 µm
Display range	Ra, Rq: 0,01 µm - 100 µm
	Ry, Rz, Rt, Rp, Rv, R3z, Rk, Rpk, Rvk, R, Rx, W, Wx,
	Wte: 0,02 µm - 350 µm Aw,
	AR: 2,0 - 350 m
	S, Sm: 2 µm - 4000 µm
	PC: 2,5 / cm - 5000 / cm
	sc: -350 µm - 350 µm (-14000 -14000 Min)
	Lo: 0,1 mm - 99,999 mm
	mr, Mr1, Mr2, mrd: 0 - 100%
	A1, A2: 0 - 15000 a, q, Ku: 0.01 - 100
	Vo: 0,0000 - 999,99
Display magnification	Vertical: 10 x, 20 x, 50 x, 100 x, 200 x, 500 x, 1000 x, 2000 x, 5000 x, 10000 x, 20000 x, 50000 x, 100000 x, AUTO
	Horizontal: 1 x, 2 x, 5 x, 10 x, 20 x, 50 x, 100 x, 200 x, 500 x, 1000 x, AUTO
Printer type	Thermal, print width 48 mm (1.9")
Statistics	Max/Min and mean value (for all parameters)
	Standard deviation(s), Pass ratio, Frequency histogram
Tolerance evaluation	Upper/lower tolerance for three parameters
Setup storage	Five setups
Auto-sleep off	After five minutes of inactivity

Appendix C

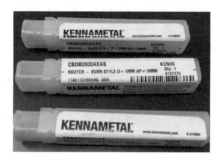

High-Performance Solid Carbide End Mills • CFRP Router

Burr-Style Router • CBDB

Features and Benefits

- Kennametal standard dimensions.
- Aerospace composites and fibreglass.

SPECIFICATIONS

Burr-Style Router • CBDB • Metric

K600	KCN05

Show 10 ▼ entries   ▼ Column Filter   Search:

order number	catalog number	Grade	D1	D	Ap1 max	L	Z U
4137486	CBDB0600AXAS	K600	6,00	6,00	18,00	63,00	12
4137475	CBDB0600AXAS	KCN05	6,00	6,00	18,00	63,00	12
4137487	CBDB0600AXBS	K600	6,00	6,00	36,00	100,00	12
4137476	CBDB0600AXBS	KCN05	6,00	6,00	36,00	100,00	12
4137488	CBDB1000AXAS	K600	10,00	10,00	18,00	83,00	12

Printed in the United States
By Bookmasters